2

WHERE PHYSICS
WENT WRONG

WHERE PHYSICS WENT WRONG

Bernard Lavenda

University of Camerino, Italy

World Scientific

NEW JERSEY · LONDON · SINGAPORE · BEIJING · SHANGHAI · HONG KONG · TAIPEI · CHENNAI

Published by

World Scientific Publishing Co. Pte. Ltd.

5 Toh Tuck Link, Singapore 596224

USA office: 27 Warren Street, Suite 401-402, Hackensack, NJ 07601

UK office: 57 Shelton Street, Covent Garden, London WC2H 9HE

Library of Congress Cataloging-in-Publication Data
Lavenda, Bernard H.
   Where physics went wrong / Bernard H. Lavenda, University of Camerino, Italy.
     pages cm
   Includes bibliographical references and index.
   ISBN 978-9814632928 (pbk. : alk. paper)
   1. Cosmic background radiation. 2. General relativity (Physics) 3. Gravity. 4. String models.
I. Title.
   QB991.C64L38 2014
   523.1--dc23
                  2014031064

British Library Cataloguing-in-Publication Data
A catalogue record for this book is available from the British Library.

Typeset by Stallion Press
Email: enquiries@stallionpress.com

Printed in Singapore by Mainland Press Pte Ltd.

To the memory of
Enzo Ferroni (1921–2007)
and
Alfonso Maria Liquori (1926–2000)

Preface

In the evening of the 17th of March 2014 the BBC reported:

> Scientists say they have extraordinary new evidence to support a Big Bang theory for the origin of the Universe.
>
> Researchers believe they have found the signal left in the sky by the super-rapid expansion of space that must have occurred just fractions of a second after everything came into being.
>
> It takes the form of a distinctive twist in the oldest light detectable with telescopes.

The date is important since the subsequent breaking news story was about the missing Malaysian jet MH 370 airliner. That airliner has been missing for almost a year now. Taken together the news briefs say we know exactly what happened 13.7 billion years ago just 10^{-34} s after it all started, but we are unable to pinpoint the position of a missing jet airliner!

What we are observing is "the oldest light detectable by telescopes, the first trillionth of a trillionth of a trillionth of a second old." Amazing! But how do we know that the photons are that old? Do they come with their birth certificates?

The telescopes might have picked up polarized radiation, but electromagnetic radiation is clearly distinct from gravitational radiation, unless someone comes up with a unified theory of electromagnetism and gravitation. Einstein couldn't do it, but that's not saying it can't be done. What we know, however, is that it hasn't been done thus far — so there is a gap in the argument.

The whole news report is therefore lopsided. Breaking news should report discoveries, whatever be their causes, and not be ammunition for prior consensus, no matter which way it will turn out.

Two months earlier, on January 26th,

Stephen Hawking [IW14] has set the world of physics back on its heels by reversing his lifetime's work and a pillar of modern physics claiming that black holes do not exist — saying that the idea of an event horizon, the invisible boundary thought to shroud every black hole (the awesome gravitational pull created by the collapse of a star will be so strong that nothing can break free including light) is flawed.

The absence of event horizons means that there are no black holes — in the sense of regimes from which light can't escape to infinity.

Hawking told *Nature*: "There is no escape from a black hole in classical theory. [But quantum theory] enables energy and information to escape from a black hole."

Resort to quantum theory was the only way of 'conserving' information, if that has any meaning at all. Similar recourse to quantum theory was also made by Einstein when after outlining with genial simplicity the theory of gravitational waves, he was led to the unacceptable result that *spontaneous* waves "should also as a rule give rise to the dispersion of energy through irradiation" [LC17]. "Since this fact" — by Einstein's own admission — "should not happen in Nature, it seems likely that quantum theory should intervene by modifying not only Maxwell's electrodynamics, but also the new theory of gravitation."

There was, in fact, no need to appeal to quantum theory — just as there is no reason in Hawking's case — for Einstein uncovered a flaw in his theory: the non-existence of a gravitational tensor which ruined his covariant formulation.

Misrepresentation by the media is only paralleled by the avalanche of popular science books that have recently hit the market, and the charlatans that utilize the media to capitalize on the public's ignorance.

Mis-selling has now become the norm rather than the exception. But it goes further than mere mis-selling; it boils down to downright untruths. In Greene's [Gre99] best-selling novel, *The Elegant Universe*, he affirms "String theory... is the story of space and time since Einstein."

Within this new framework, general relativity and quantum mechanics *require one another* for a theory to make sense. According to superstring theory, the marriage of the laws of the large and small is not only happy but inevitable... String theory has the potential to show that all of the wondrous happenings in the universe — from the frantic dance of subatomic quarks to the steady waltz of orbiting binary stars, from the primordial fireball of the big bang to the majestic swirl of heavenly galaxies — are reflections of one grand physical principle, one master equation.

Not only does he not tell us what that equation is, he does not even exhibit the potentialities of superstring theory. This is not from any oversight, but simply from the fact that he can't. The theory has never made any refutable predictions, nor will it ever.

Five years later, and a lot of publicity in between, we find Greene [Gre04] promising us much of the same and more:

> In our own era we encounter inflationary's gratifying insights into time's arrow, string theory's rich assortment of extra spatial dimensions, M-theory's radical suggestion that space we inhabit may be but a sliver floating in a grandiose cosmos, and the wild speculation that the universe we see may be nothing more than a cosmic hologram.

Greene fails to mention eternal inflation which wouldn't jive with time's arrow, and fails to define what M-theory is, because he can't. He later admits that "M being a tantalizing placeholder whose meaning — Master? Majestic? Mother? Magic? Matrix?—awaits the outcome of a vigorous worldwide research effort now seeking to complete the new vision illuminated by Witten's powerful insight." That insight is more than twenty years ago, and has gone stale; nothing has happened since then, and I'm confident that it won't be completed. Maybe it stands for Messiah? — whose arrival is anxiously being awaited.

This is not science but some primitive form of god-worship. These accusations are not due to "an author [having] a lot less control than you think over things like book title, cover design and jacket copy," [Bag13] because no publisher in his right mind would ever put on the dust jacket an assertion that the author of a book would disagree with or find distasteful. This is willful and premeditated misrepresentation of a theory that doesn't preform to what it has been chalked up to be.

Baggott raises the question of whether authors of string theory books believe what they are writing about.

> When asked directly at the Isaac Asimov Memorial Debate at the American Museum of Natural History in New York City, which took place on 7 March 2011, Greene replied: 'If you asked me, "Do I believe in string theory?", my answer today would be the same as it was ten years ago: No.'

They are out on a limb, much too far to come back so that there is only one way to go — down. It has been a total waste of money, time and, above all, human life. Physicists could have dedicated themselves to something nobler than following the pack. And publicly funded organizations, like NASA, could have been a lot less dogmatic and more impartial until all the results are in. The standard model is just that—a model—and not a credible one at that.

Baggot recriminates himself for the reasons of writing his book. But his blurb is quite to the point:

> Scientists are suffering from a 'Grand Delusion'—a belief that they can describe reality using mathematics alone, with no foundation in scientific fact. The result is 'fairy tale' physics.
>
> A string [no pun intended] of recent best-selling popular science books has helped to create the impression that fairy tale physics is established science.

A noble cause if there ever was one, but has he provided the "needed antidote"? I'm sorry to say that he hasn't because he never once crosses swords with the feeble and inaccurate foundations upon which these theories seemingly sprout from. General relativity tops the list which is hardly 'general' and even less 'relativistic.' It is this task that I will undertake in the following pages.

I wish to express my grateful thanks to Giancarlo Nardini, librarian of the University of Camerino, for procuring articles that were not available on the Internet. I would also like to thank Daniel Kuster for his patience in helping me wade through all the problems I had with VerbTeX Pro.

Bibliography

[Bag13] J. Baggott. *Farewell to Reality: How Fairytale Physics Betrays the Search for Scientific Truth.* Constable & Robinson Ltd, London, 2013.

[Gre99] B. Greene. *The Elegant Universe.* Jonathan Cape, London, 1999.

[Gre04] B. Greene. *The Fabric of the Cosmos: Space, Time and the Texture of Reality.* Alfred A. Knopf, New York, 2004.

[IW14] Stephen Hawking: "The Absence of Event Horizons Means There Are No Black Holes". *Science*, January 26, 2014.

[LC17] T. Levi-Civita. On the analytic expression that must be given to the gravitational tensor in Einstein's theory. *Rendiconti della Reale Accademia dei Lincei*, 26:381, 1917.

Contents

Introduction

In his popular science book, *Thirty Years that Shook Physics*, Gamow divided the two great revolutionary theories that changed the face of physics during the early decades of the twentieth century into quantum theory and relativity. The former being the creative work of groups of physicists, while the latter was the work of a single individual. The title of the book refers to quantum theory, whereas relativity, which was a side show of a side show, became the real watershed that transformed present day physics into the mess that it's in.

Although there has been a backlash against popular science books that mislead the general public into believing that it is only a matter of time that superstring theory and multiverse universes will open up a whole new reality to explore, no one dares touch the precarious, if not entirely wrong, tenets upon which these fanciful theories are based. This is a consequence of the fact that the people who are writing the critiques are not physicists themselves, or if they are, not active and critical ones. However, it isn't possible to sever the limbs without cutting the trunk.

The single individual who created general relativity is Einstein, with a little more than a bit of help from Poincaré. And not only did Einstein create general relativity, he also created his own cult that now pervades almost all of physics. The Einstein myth is that new physics does not need new experiments, and can be replaced by thought, or *gedanken*, experiments. It has been the rallying cry of string theorists who see it as the way of completing Einstein's dream of a unified theory without getting bogged down in details like experimental verification, even though there is a lot more to unify now than in Einstein's day.

In a most scathing attack on Einstein and his science, Schlafly [Sch11] finds it necessary to detract, rather than criticize what can be considered as Einstein's sole scientific offspring: general relativity which is not a generalization of his special relativity since it contains no relativity at all.

Isaacson [Isa09] tells us that the general theory of relativity "was the product of a decade of solitary persistence during which Einstein wove together the laws of space, time, and motion." To this Schlafly retorts: "No, it was not solitary. It was a product of mathematicians telling Einstein how to construct a nonlinear relativity theory." Implicit in this statement is that what the mathematicians were telling Einstein was correct.

No, the mathematicians did not tell him to construct an indefinite metric that implies gravity propagates at the speed of light. (The propagation speed is what Poincaré told him as well as the possibility of the existence of gravitational waves.) One telling indication that things are not what they appear to be is that when the spatial components of the metrics used in general relativity are constrained by constant curvature they do not coincide with the known constant curvature metrics of non Euclidean geometries. Moreover there is nothing unique to Einstein's explanation of the advance of the perihelion of Mercury, the gravitational deflection of light, or the gravitational shift of spectral lines.

Eddington [Edd29] remarked early after the appearance of general relativity "any problem on other paths of rays near the sun can now be handled by the methods of geometrical optics applied to the equivalent refracting medium." However this is not quite true since you need one approximation beyond the geometrical optics limit: the short wave diffraction limit where the relativistic phenomena appear as small corrections to what would otherwise be rays of geometrical optics [Lav11].

Furthermore none of the tests of general relativity require the full indefinite metric: the advance of the perihelion and gravitational deflection require only the space part of the metric while the gravitational shift of the spectral lines is accounted for in the coefficient of the time component of the metric. Regarding general relativity in general, there is nothing beautiful or aesthetically pleasing of having a gravitational tensor which is *not* a tensor at all, and conservation laws which can hardly be considered as such [cf. §3.5]. The lack of a gravitational tensor raises havoc with Einstein's putative principle of general covariance.

And Einstein's principle of equivalence between acceleration and gravitation can hardly be called a principle. It is not the equivalence between gravitational and inertial mass that is at stake here, as Schlafly is quick to recall is "conventional wisdom going back to Galileo and Newton," but, rather, that gravitational acceleration can be neutralized by any other form of acceleration, say by turning on the rockets in a free-falling elevator.

As another typical example, Schlafly asks us to consider:

> A simple thought experiment shows that acceleration leads to curvature. Consider a simple disc. Euclidean geometry teaches that the circumference is 2π times the radius. Now spin the disc very fast. Special relativity now teaches that the circumference has a length contraction, while the radius remains is unchanged. The circumference is not 2π times the radius anymore. The disc is curved.

Special relativity cannot make any pronouncements on curvature since accelerations are involved. Hence, the length contraction due to uniform motion is a *non sequitur*. Einstein rationalized the opposite result that rulers laid tangentially along the circumference would undergo contraction and therefore you would need more of them to measure the circumference when the disc is in motion than when it is at rest. Hence the ratio of the circumference to the radius should be greater than 2π. The answer is correct but the reasoning is faulty [Lav11].

Instead a non-Euclidean metric of constant negative curvature is involved in which the coefficient of the spherical component of the metric is the square of the circumference of a hyperbolic circle, when multiplied by $4\pi^2$. This result was already known to Gauss [Lav11]. Hence, the circumference is larger than the Euclidean circumference, and not smaller than due to length contraction, as Einstein surmised but for the wrong reason.

Where is gravity in the spinning disc? And how can it neutralize centripetal acceleration? It is not in the metric, and has to be appended on when light propagates through a medium of varying index of refraction. This was appreciated by the young James Clerk Maxwell in the 'fish eye' that was named after him [Lav11]. It is completely extraneous to a uniformly rotating disc, and therefore, Einstein's equivalence principle is hardly a principle.

The lesson to be learned from this example is that Einstein's friend Grosmann should have alerted him to the fact that there was something called non-Euclidean geometries of constant curvature, and the spatial metrics that were derivable from his equations had to reduce the same metrics in the limit of constant curvature. How can the famed Robertson-Walker metric, which is used in the standard model, be relevant to an expanding universe when it confuses the scale factor with the circumference of a non-Euclidean circle [Lav11]? Castles cannot be built in on quicksand! How can any faith be put into results which follow from faulty premises? A prime example is inflationary cosmology.

Even disregarding all these drawbacks of Einstein's general relativity, we can ask ourselves: what is the sense of modeling the violent universe as an ideal fluid? How can such a description apply to the Crab Nebula, believed to be the remnant left when a star exploded many centuries ago, or of supernova explosions that blow apart entire stars? Or what about globular clusters, quasi-stellar objects, known as quasars, which are star-like in appearance, but much more massive than stars, with odd spectral properties that seem to make them much further away than they really are [Arp98]? And last but not least, the big bang itself which would defy the laws of statistical mechanics and thermodynamics if it started off in a state of thermal equilibrium, which inflationary cosmology predicts, and somehow evolved into a less likely, non-equilibrium, state. This would upset both our belief in the second law and time's arrow. It can hardly be considered to be the behavior of an ideal fluid on any scale!

The standard model predicts that the big bang was both homogeneous and isotropic — indeed an odd way for an explosion to occur. Supposedly these properties are well-described in the Robertson-Walker metric. So how can we look to such metrics for information concerning the exponential growth of the universe almost immediately after birth? The answer is simple: the equations of state that predict a negative, expansive pressure are wrong. Both inflation and the de Sitter universe suffer from the same disease.

Then at recombination, which is a misnomer because electrons had not combined with baryons previously, the gases would have to be in thermal equilibrium that would imply perfect black body radiation. This is indeed astonishing since it would try to describe the life of an individual beginning with his burial! As Boltzmann has so rightly insisted, systems evolve from states of lower probability to states of higher probability, and there can be no higher state than that of a black body. Boltzmann's H-theorem clearly shows that thermal equilibrium is the place of arrival — not the place of departure. It has often been claimed that theories competing with the big bang cannot explain the cosmic microwave background radiation (CMBR) with such naturalness and elegance. The big bang, however, must have the film running backwards!

At recombination, the temperature of the universe would have been around 3000 K and if the radiation were that of a black body, the peak in the spectrum would be at 1μm, as determined by Wien's displacement law, $h\nu_{\max}/T = 2.822$, where $\nu_{\max}$ is the frequency where the radiation peaks, and T is measured in energy units. Since the universe was in expansion, those photons would have undergone a red shift in frequency of around 10^3 since then. An increase in the wavelength by such a factor would displace the peak of the radiation curve to about 1mm, placing it squarely in the microwave region.

There is no doubt that Wien's displacement law played a pivotal role in Planck's derivation of his spectral law. The derivations of this law was based on the Doppler shift of light rays bouncing off of moving mirrors [Lav91]. But this had nothing to do with black body radiation itself. It is to the clarification of Wien's law directly in terms of the electrodynamic modes in the cavity that Ehrenfest addressed himself in 1911.

The key to Ehrenfest's derivation was to be found in a 1902 paper by Lord Rayleigh on the radiation pressure of vibrating systems in general, no matter what their origin might be. It was a statement on the change in energy of a normal mode on *sufficiently* slow variations in the dimensions of the enclosure. Any sufficiently slow, i.e., 'adiabatic,' variation in the reflecting walls of the cavity would leave the ratio of the energy of the normal mode to the frequency of vibration of that mode invariant, the latter varying inversely to the linear dimension of the cavity. It was crucial that the motion of the reflecting walls be slow in order that the normal mode is not destroyed, and this required a great number of oscillations to occur in any finite displacement of the walls of the cavity. In thermodynamic jargon, the process had to be carried out 'quasi-statically'.

Since the expansion of the universe cannot be considered as a quasi-static process, it is difficult to envisage how the black body radiation retained its thermal character. But this is not what Penzias and Wilson observed. They observed the radiation at 7.35 cm only because it was this wavelength that Bell Labs had been using to communicate with orbiting satellites. This wavelength was many times longer than the wavelength at which the spectrum should peak if it was in fact black body radiation at all. But, that was a mere conjecture at the time, and Wien's law applied to the 7.35 cm wavelength should have led to a lower temperature than that determined by the true black body curve.

Since terrestrial measurements could not measure microwaves with wavelengths less than 3 cm because of water absorption, confirmation had to await over a quarter of a century until COBE (Cosmic Background Explorer) was launched. Measurements in wavelength were made over a range from 0.1 to 10 mm. COBE confirmed a perfect black body spectrum at the predicted temperature of 2.725 ± 0.002 K.

Such a confirmation could hardly be based on Rayleigh's theorem concerning the nature of radiation pressure under the quasi-static variations of the reflecting walls of the enclosure. What enclosure?

Hot bodies radiate electromagnetic waves so that some form of electromagnetic signal should be detected as those measured on wires. In the early decades of the twentieth century, Johnson observed electrical fluctuations caused by heat and described them in terms of a fluctuating voltage across a resistor rather than in terms of electromagnetic waves. Nyquist applied the principles of statistical mechanics to describe such a phenomenon and related the mean square voltage of the noise created by heat to the product of the resistance of the resistor, the thermal energy and the band width that depends on the properties of the measuring device. Another quarter of a century was to pass until Nyquist's formula was to be enshrined in the Callen-Welton fluctuation-dissipation theorem of linear irreversible thermodynamics, which relieved it of its specificity.

What does Johnson noise, whose power is the product of the thermal energy and the band width, have to do with a black body radiation spectrum? To convert it into the spectral density of black body radiation in the low frequency limit, known as the Rayleigh-Jean's limit, it is necessary to multiply the noise power by the absorption cross-section, which is the area that a circle whose circumference is the wavelength of the radiation would occupy.

When the CMBR was found to be isotropic, there was a sigh of relief that it could be accounted for by the big bang cosmology. Then there was the afterthought that isotropy would not lead to discrete structure formation so the search was on for anisotropies. Penzias and Wilson originally set the anisotropy level to be about 10%. Simplistic assumptions regarding gravitational effects set the level of fluctuations at 0.1–1%. When it became evident that no such anistropies, interpreted as relative temperature fluctuations, $\Delta T/T$, were to be found, the level had to be lowered to $\Delta T/T \sim 10^{-4}$. Again no such fluctuations were to be found. But when inflation intervened in the 80's, which required the universe to evolve at the critical density, with the need to postulate the presence of dark matter to fill up the gap of missing mass, the fluctuation level could easily be rationalized to be lower than that of 10^{-4}. This again was compatible with observation — or better yet the lack of observation!

It was insisted that the light elements were synthesized in the big bang, and not inside stars, and the low level of fluctuations needed dark matter/energy to be invented so that the critical density could be achieved for inflation to take over. COBE first began signaling anistropies at the large angular scale of $\sim 10°$ in 1992 which registered a relative thermal fluctuation of the order of 10^{-5}, resulting in a temperature difference of $30 \pm 5\ \mu K$.

Such measurements were greeted with wild enthusiasm claiming the Holy Grail had finally been found, and the face of God had been sited. Taking over from COBE, WMAP (Wilkinson Microwave Anisotropy Probe) was especially designed to measure asymmetric patterns in CMBR. The fluctuations were required, and confirmed, to have a Gaussian spectrum. Now behind the Gaussian assumption lies the law of large numbers, and it is hardly what is needed to create the seeds for galaxy formation. Gaussian fluctuations are small by definition; they are independent and identically distributed thereby excluding any (nonlinear) correlations that would be essential to structure formation.

Background radiation comes in all sizes and shapes. The sole exception appears to be the CMBR which is exclusively reserved for microwaves. Starlight arrives at us from the absorption and re-radiation by dust so why not the CMBR?

If gravity propagates at the speed of light, as general relativity assumes, then why don't we see gravitational aberration? Also a finite speed of propagation would result in the distortion of planetary orbits which is not observed [Fla93]. We are at the doorstep of creating a theory of 'everything,' but we don't know what everything is or what the basics are. We are told that space and time has been amalgamated into a single four-dimensional structure only to be told that space itself can expand at superluminal velocities during a period of inflation. And since gravitational waves are distortions of space, stretching in one direction and squeezing in the perpendicular direction, while always being normal to the direction of propagation, begs the question: why don't gravitational waves propagate faster than the speed of light when space itself can?

Gravitational waves are supposedly created whenever large masses move rapidly. So gravitational waves would have been generated in the early universe. According to Maxwell, accelerated charges produce electromagnetic waves, while, according to Einstein, accelerated masses produce gravity waves [Web12]. However, according to Einstein's equivalence principle accelerations can be compensated, and annihilated, so that the motion would disappear, and so too, the gravitational waves!

Inflation would have gravitational waves stretched in space. But where did these large masses come from in the early universe? There were not even baryons present at that time, let alone large masses that could generate gravitational waves. Notwithstanding this, gravitational waves could propagate more freely than electromagnet waves before the period of recombination because it would not be influenced by the fog comprised of leptons and baryons. Thus, when the universe became transparent after recombination, the gravitational waves were able to leave their imprint on the CMBR. How is it possible to leave an imprint on something like radiation that is not solid? This is tantamount to trying to explain the working of a steam engine on the shape of the coals that stoke the fire in the boiler [FHN00]. It is completely superfluous.

Electromagnetic waves can become polarized when distorted in space, but so can gravitational waves. The distinction lies in what are the types of patterns that should be expected around hot and cold spots in the CMBR. But what do temperature inhomogeneities have to do with the polarization of electromagnetic waves?

Whereas electromagnetic waves should show E-mode polarization, or gradient-mode polarization, gravitational waves should manifest B-mode polarization which would be a vortex-type of motion. The same promoters of the E- and B-modes claim that electromagnetic waves are polarized at 90°, while gravitational waves are polarized at 45°. Since electromagnetic waves can be both linearly, elliptically, or circularly polarized this distinction makes no sense. Moreover electromagnetic waves have a different mechanism of propagation than the hypothetical gravitational waves so there is no common ground for comparison. In other words, unlike electromagnetic waves that carry their electric and magnetic fields piggyback, gravitational waves have no such analogous fields, although attempts have been made to attribute them with such types of fields [Lav11].

The strength of the B-mode is an indication of how big the gravitational waves were. According to inflation, the strength of gravitational waves depends on the magnitude of their wavelength: "the longer the wavelength the stronger the gravitational waves" [Web12]. This seems queer since the energy is proportional to the frequency which is inversely proportional to the wavelength. The strength of the gravitational waves depends also on how rapidly the universe expanded, which in turn, depends on the precise moment that inflation began. Here were are talking about times of the order of 10^{-34} s. If large enough, gravitational waves are expected to be 'imprinted' on the polarization of the CMBR. This is probably the first time anybody ever spoke about imprinting something on the polarization of electromagnetic radiation.

It is commonly believed among cosmologists that the universe is expanding, although there are those that would beg to differ [Rat10, Ler91]. Since the horizon is limited by the velocity of light, while the smoothing of space by inflation is not, the actual cosmos is much greater than the patch we can observe. It is not to be discarded that other patches underwent their own inflation that would give rise to other universes with properties dissimilar from our own with different physical laws. Here we are on the verge of science fiction, and anything goes.

And science fiction it is: for here is where inflation joined hands with string theory. This is indeed a first rate example of the blind leading the blind. The apparent oddity of there being eternal inflation joined forces with the absence of a preferential vacuum state of superstring theory. A simple count sets the genus, or number of holes in the Calabi-Yau shapes, at a mere 10^{500}. This represents the number of ways that the extra six dimensions of superstring theory can be compactified, or rolled up. For it was the 'success' of inflation and multitude of vacuums, or valleys in the 'Landscape' that made the 'Anthropic' principle appealing [Sus06]. The Anthropic principle makes the dumb assertion that physical laws have to be those, otherwise we would not be here.

The landscape is not that of a gardener, but rather, represents the number of different ways of compactifying the unwanted dimensions. Each different way of achieving this corresponds to a different type of universe with different properties and physical laws. In the language of inflation, each bubble represents a different universe, and our universe is found among the multitude.

Such a theory must not be taken seriously, as Susskind has. But Susskind doesn't deal with physical principles and measures success in the number of converts he can count. He is undoubtedly the most 'democratic' physicist of our time, setting public (peer) opinion above simple common sense. But, sadly for him, science in general, and physics in particular, is not a democratic institution, and it's not the number of votes that counts, or the number of friends you have on the editorial board that should determine scientific results!

The exponentially large number of vacuum environments makes fine-tuning in the early universe feasible. The fine balancing act enters when the very big, i.e., gravity, meets the very small, i.e., elementary particles. Attention is focused on, for lack of a better choice, Einstein's nonsensical fudge factor, his so-called 'cosmological constant'. It represents a repulsive force that counteracts the self-destructive force of gravity. In order that our universe appears the way it is, it would have to be incredibly fine-tuned. But a constant is a constant, and as such it is impervious to changing conditions.

The repulsive, constant, cosmological constant acts like a negative pressure. We all know what positive pressure does, it makes things expand. So, negative pressure, by definition, must make things contract, right? Wrong! Consider a cylinder fitted with a piston. Drawing the piston out, increases the vacuum so that there would be an undeniable tendency for the piston to contract when left to itself. But, because you increase the vacuum as you pull the piston out, there is more vacuum energy around, with density constant, and this must be associated with a negative pressure. For if the pressure were positive it would tend to push the piston out.

But by its very definition, a vacuum is devoid of energy and matter. However this is not what is implied by general relativity. A classic example is the expansion of de Sitter's empty universe. When pressed by the question of "what makes the universe expand or swell up?", de Sitter raised his hands in despair and answered, "This is done by Lambda. Another answer cannot be given." Lambda is Einstein's cosmological constant.

It is this sort of contorted and twisted reasoning that pervades Einstein's theory of general relativity in which pressure, along with mass and energy, contributes to gravitational effects. Einstein taught us the equivalence of mass and energy, but pressure?[1] Since the cosmological constant has precisely the negative pressure required to balance the positive energy density, it is the cosmologist's favorite explanation for the observed acceleration of the universe.

However, all known gravitational effects are attractive. The error is in attributing energy to something that has none, i.e., the vacuum. Pulling the piston out requires energy, and letting it go will make it return to its

---

[1] Again the equivalence of the change in energy and mass is due to Poincaré who derived it, at least five years before Einstein did, from the recoil of an artillery piece when fired [Lav11].

original position. You can say that negative energy has made the volume contract to its original value, in contradiction of what the cosmologists claim. Elementary physics, my dear Watson!

The problems raised here, like the 10^{500} vacuum states from which to choose from, give the illusion of a designed universe. It is like the illusory bone of contention of whether information is lost when it is flushed down a black hole. Neither Susskind nor his arch opponent, Hawking, bother to define what is meant by information, except from some loose references to the number of 'bytes,' or to the meaning of entropy. Glaring are the inaccuracies in Susskind's [Sus08] arguments like: (i) "the existence of entropy doesn't imply that a system has a temperature," (ii) black hole evaporation, where black holes shrink by radiation until they are no larger than an elementary particle, and then in a puff they are gone, and (iii) "the maximum amount of information that can be possibly contained in any region of space cannot be greater than what can be stored on the boundary of a region, using no more than one-quarter of a bit per Planck area."

Regarding (i), how do you define temperature if not through the variation of the internal energy with respect to entropy? Regarding (ii), black hole evaporation has been compared to the evaporation of a puddle on a hot summer's day by Sciama. However, puddles don't get hotter the smaller they become, and don't use black body radiation as the mechanism of their evaporation. Finally, (iii) is a reference to the so-called holographic principle, introduced needlessly by 't Hooft, who was working under the erroneous impression that the surface area of a system has something to do with its entropy.

We can, again, trace back this sort of sterile polemics to Einstein who reasoned that all major scientific breakthroughs will come from pure reason alone without the intervention of experiment. Einstein can be accredited with two major mechanisms for avoiding experimental confrontation: the *gedanken* experiment and the presence of imaginary observers. The former includes the desired conclusion in its premise while the latter can negate physical facts.

An example of the latter would be the comparison of observers clocks which are placed at different points on a rotating disc. According to Einstein, the watch of an observer who is on the rim of a uniformly rotating disc would appear to go slower than an identical watch worn by the observer located at the center of the disc. However this is, indeed, false because no communication between the observers would ever reveal their positions on the disc, because the disc belongs to hyperbolic space, and not Euclidean space [Lav11].

So it appears that not only Einstein's theory of general relativity is at the root of the present dilemma physics is faced with, but also his *deus ex machina* for arriving at his preconceived conclusions.

Bibliography

[Arp98]   H. Arp. *Seeing Red: Redshifts, Cosmology and Academic Science.* Apeiron, Montreal, 1998.

[Edd29]   A. Eddington. *Space, Time and Gravitation: An outline of the general theory of relativity.* Cambridge University Press, Cambridge, 1929.

[FHN00]   G. Burbidge F. Hoyle and J. V. Narlikar. *A Different Approach to Cosmology.* Cambridge University Press, Cambridge, 2000.

[Fla93]   T. Van Flandern. *Dark Master, Missing Planets and New Comets.* North Atlantic Books, Berkeley, CA, 1993.

[Isa09]   W. Isaacson. *Einstein's Final Quest. In Character,* Spring 2009.

[Lav91]   B. H. Lavenda. *Statistical Physics: A probabilistic approach.* Wiley-Interscience, New York, 1991.

[Lav11]   B. H. Lavenda. *A New Perspective on Relativity: An odyssey in non-Euclidean geometries.* World Scientific, Singapore, 2011.

[Ler91]   E. J. Lerner. *The Big Bang Never Happened.* Random House, New York, 1991.

[Rat10]   H. Ratcliffe. *The Static Universe.* Apeiron, Montreal, 2010.

[Sch11]   R. Schlafly. Dark Buzz, www.DarkBuzz.com, 2011.

[Sus06]   L. Susskind. *The Cosmic Landscape.* Little, Brown and Co., New York, 2006.

[Sus08]   L. Susskind. *The Black Hole War: My battle with Stephen Hawking to make the world safe for quantum mechanics.* Little, Brown & Co., New York, 2008.

[Web12]   S. Webb. *New Eyes on the Universe.* Springer, New York, 2012.

Chapter 1

The Unobservable Universe

1.1 All What's Not Known about CMBR

1.1.1 *Predecessor of CMBR*

According to Goldberg and Scadron [GS81] "The greatest blow against steady state theory was struck in 1964 with the accidental discovery of what has since been interpreted as cosmic microwave background radiation." This is supposedly the radiation left over from the big bang. It is isotropic and its spectrum follows a black body curve at a temperature of 2.7 K.

Why it supposedly sounded the death knell of the steady state theory was that, according to steady state theory, there is never enough matter around that would allow it to be in equilibrium with radiation. Hence, there was no possibility for *bremsstrahlung* (breaking) radiation that would eventually lead to a unique black body spectrum.

Steady state theorists would beg to differ [FHN00]. Even before the detection of the CMBR by Penzias and Wilson, the two big schools of cosmology at that time, the Russian school led by Zel'dovich and the American one headed by Dicke, were convinced that the hot big bang scenario was correct, and the 'smoking gun' of the leftover radiation from the bang would be found. The prediction of such a radiation from steady state theory would be highly contrived to say the least.

There was mounting already a considerable body of evidence that something similar to the CMBR was present from the 1940's observations of the presence of interstellar diatomic molecules like CN, CH, and NaH. According to McKellar [McK40], "If these identifications are proved true, they are of considerable interest and importance in that they constitute the first definite evidence of molecules in interstellar space." McKellar even went so far as to associate a 'rotational' temperature to the radiative excitation of the line $R(1)$ from the $R(0)$ level, cautioning that "the concept of such a temperature in a region with so low a density of both matter and radiation has any meaning."

McKellar expressed the intensity of the band line in terms of a Boltzmann factor,

$$I = I'e^{-E_r/kT},$$

where I' is the intensity factor, and E_r is the rotational energy of excitation. Almost prophetically, McKellar concluded that if $R(1)$ is not more than one-third as intense as $R(0)$ the maximum 'effective' temperature of interstellar space would be 2.7 K!

If McKellar had connected the rotational transitions to the type of radiation they produced then the microwave radiation could have already been predicted in the 1940's — a quarter of a century before it was actually found. Moreover the type of absorbers would be particles of size 1 nm, and the radiation would be of wavelengths in the range $10^6 - 10^7$ nm.

Due to its timing with the onset of World War II, and the obscurity of the journal, McKellar's paper was all but forgotten. Shortly after the war Gamow and associates, Alpher and Herman, pointed out that if helium was to have been synthesized in the early universe, and not in stars like the sun as was commonly believed, there should be a present day interstellar radiation field with a temperature of 5 K.

Gamow assumed that the baryon-photon ratio, $\eta \cdot 10^{-10}$, has remained constant from the time of helium synthesis in the early universe to the present day. If the mass fraction of helium Y was one-quarter, a value that was commonly quoted for the mass fraction in stars at that time, then $\eta \sim 4$. Moreover if η_b represents the baryon number density, then its ratio to the photon density of black body radiation, $20.3T^3$ cm^{-3}, would be given by

$$\frac{\eta_b}{20.3T^3} = 4 \cdot 10^{-10}. \qquad (1.1)$$

Assuming that the present day density of baryons is 10^{-30} g/cm^3, (1.1) gives a temperature of 5 K.

Rather, if we plug in $T = 2.73$ K into (1.1) with $\eta = 6$, out pops $\eta_b = 2.51 \times 10^{-7}$ cm^{-3}. Multiplying this by the mass of a proton gives the present day mass density of baryons as 4.2×10^{-31} g/cm^3, or 0.062 of the critical mass density, $3H^2/8\pi G$, if the Hubble constant is taken as $H = 60$ Mpc. This is the so-called 'missing mass' problem for which it argued that the remaining mass must be 'dark' mass-energy.

1.1.2 *Is the CMB Black Body Radiation or Noise?*

In his book *The First Three Minutes* Weinberg contends:

> The detection of the cosmic microwave background radiation in 1965 was one of the most important scientific discoveries of the twentieth century. Why did it have to be made by accident? Or to put it another way, why was there no systematic search for the radiation years before 1965?

Nothing could be further from the truth, and yet Weinberg offers three reasons why it should be true:

1. The success in explaining the abundance of the elements through stellar nucleosynthesis made the need for primordial nucleosynthesis superfluous which was supporting a steady state theory.
2. There was no communication between theorists and astronomers so that the latter were ignorant of the possibility of there being thermal radiation at a very low temperature.
3. Physicists prior to 1965 did not take cosmology seriously.

This is certainly an inexact historical account of what really happened. What really transpired was this.

In May 1964, the Americans, Penzias and Wilson, attempted to fit a 6 meter radio antenna in Holmdel NJ for use in astronomy. After carefully removing all the noise they could, there still persisted a weak signal which was independent of the direction the antenna was pointing in.

At about the same time, the Russians, Doroshkevich and Norikov, explored the possibility of measuring the 'relic' radiation that was predicted some two decades earlier by Gamow. In their article they refer to the report of Ohm in 1961, also of Bell Labs, in which he refers to a strange excess noise found in the same 6 meter antenna, which he converted into a 3 K temperature. Ohm did not pay much attention to this noise because he thought it was well within experimental error.

Only a few miles away from where Penzias and Wilson were working, Dicke and Peebles were attempting to build a suitable antenna to detect the same relic radiation. In January 1965, Penzias tells Burke of MIT his frustration in not being able to get rid of the persisted noise and the latter suggests that Wilson and he get in touch with Dicke, who after speaking to Penzias by phone tells his group, "Boys, we've been scooped!" The irony of it all is that Penzias and Wilson used the same microwave radiometer that was first devised by Dicke some two decades before.

In May 1965 Penzias and Wilson announced in the *Astrophysical Journal* the excess antenna radiation of 3.5 K, and in an accompanying paper, Dicke, Peebles, Roll and Wilkinson refer to the photons as coming from the 'primeval fireball.' This is a far cry from Weinberg's tale of how the CMBR was discovered, but nevertheless emphasizes how facts are twisted to suit individual needs.

Dicke [Dic46] explains the seemingly paradoxical conclusion that black body radiation, in the low frequency Rayleigh-Jeans limit, is proportional to the square of the frequency whereas Johnson thermal noise is frequency independent. The power of the antenna, proportional to the average absorption cross section, goes as the square of the wavelength. That is, if the distance to the antenna is a certain number of its wavelength then the cross section of the shadow zone, where the circular waves emitted by the antenna are close enough to the incoming plane waves from a distant transmitter, permit them to be in phase with each other.

In other words, an antenna becomes a receiver by becoming a transmitter. The square of the wavelength of the antenna precisely cancels the square of the frequency in the density of states in black body radiation making the absorbed power completely independent of the frequency, just as in Johnson thermal noise.

Dicke's set-up consisted of an antenna pointed into a black body cavity which was connected by a coaxial cable to a resistor. Thermal radiation that is emitted by the walls of the encompassing cavity is picked up by the antenna and is transmitted down the line to the resistor where it is absorbed. Alternatively, Johnson noise in the resistor causes a noise power to be transmitted in the opposite direction which, passing through the antenna, is absorbed by the walls of the cavity.

Different power transmissions would result in different temperatures so that the resistor would either lose or gain energy. This is precisely outlawed by the second law of thermodynamics, and consequently, the available Johnson noise power from the resistor is exactly equal to the power picked up by the antenna and pointed to the walls of the cavity at the same temperature, and vice versa.

However, the dimensions of the two systems are not the same. Whereas Johnson noise is one-dimensional, cavity radiation is three-dimensional, and consequently, the density of states are different meaning different frequency dependencies. Again we are back to the situation where the power of Johnson noise is frequency independent, while in black body cavity radiation it is proportional to the square of the frequency. And again, the absorption cross section of the antenna should even the score.

Nevertheless, the antenna can be replaced by a speck of charcoal which is capable of absorbing and emitting radiation at all frequencies without disturbing the power balance. The power balance, envisioned by the second law, should not depend on the specific nature of the resistor or the antenna. Alternatively, if it did depend on the antenna characteristics, such an arrangement would not be able to pick up black body radiation, as was done by the scanning radiometer aboard the COBE satellite which obtained a perfect black body spectrum at a temperature of 2.728 ± 0.002 K, measured over a wavelength range of $1-10$ mm.

Disregarding the fact that black body radiation should be independent of both the composition of the cavity and the way it is measured, Dicke's contention would imply that above the critical frequency where equipartition, or the Rayleigh-Jeans limit, has been exceeded, the spectral density, according to Nyquist, would be proportional to $h\nu$ times the Planck factor, $1/(e^{h\nu/T} - 1)$, which when integrated over all frequencies would lead to a T^2-law for the energy density, and not a T^4-law, as required by Stefan's law.

There are some really embarrassing questions that spring to mind when forcing oneself to accept the party line that CMBR is the afterglow of the big bang. Black body radiation needs an isothermal cavity in which to establish a dynamic equilibrium between absorption and emission of radiation at all frequencies. Einstein was well aware of such a dynamic equilibrium as far back as 1917, where he predicted a new, non-classical emission in the form of stimulated emission.

But where is the cavity in the CMBR? If the frequencies have been red-shifted then according to Wien's displacement law, the temperature corresponding to where the spectral density peaks will also be red-shifted by the same amount. This cannot explain the exact black body spectrum at $T_m = T/(1 + z)$ at frequencies out to centimeter wavelengths that are obviously different from ν_m. And if the red-shift, z, is invoked in the black body spectrum wouldn't it also hold in McKellar's determination of the

5 K, or the excitation temperature of the ground-state of the CN molecule, making it thousands of degrees greater?

Provided the spectrum was not distorted by the addition of energy after the epoch of recombination, which is hardly feasible in an adiabatic expansion, the big bang people maintain that the hot bang produced an exact thermal spectrum. This spectrum is supposedly preserved even though the universe expanded in volume by a (mere) factor of 10^{19}, as well as being a million times hotter than it is today and 10^{19} times denser. This supposedly justifies their claim that spectral measurements of the CMBR can probe the physical conditions of the universe back to an epoch a few weeks or months after the bang.

It also supposes that nothing transpires in between. It is as if we bore a tiny hole into the black body cavity but instead of placing the bolometer at the hole, we move it a few kilometers from the hole and claim that it would have no effect on the measurements. Yet, it is known that starlight will heat a black body to a very low temperature of 3 K when it is located far from the star. Radiation by warm dust particles, either in the galaxies or intergalactic spaces, might have produced the CMBR.

Moreover, if the CMBR is the afterglow of the hot big bang, how did the universe start out from a state of thermal equilibrium, which is usually the end state to which systems evolve? Why should the photons reaching us, which have been emitted after recombination has occurred, retain their memory of black body radiation?

In this regard, neither spontaneous nor induced Thomson scattering will be sufficient to maintain a black body spectrum, contrary to what Rees [Ree68] has claimed. In plain language, photons must be 'eaten' and 'regurgitated' in order to establish black body radiation, and not simply be scattered reversibly at the same frequency, as in Thomson scattering.

Another worrying statement by Rees is that "the spectrum of background radiation is surprisingly sensitive to anisotropies in the Universe." Black body radiation is impervious to anisotropies which can only act to destroy it. And, last but not least, there is the question of how the black body radiation could have survived the extremely rapid adiabatic expansion of the universe.

Ehrenfest's adiabatic theorem requires the walls of the cavity to move ever so slowly as to not disrupt black body radiation. Ehrenfest's theorem was distilled from an earlier observation by Lord Rayleigh that if the walls of the enclosure are made to approach each other infinitely slowly, the energies of the vibrations, E_ν, change in exactly the same proportion as the frequencies, ν, so as to render their ratio, E_ν/ν, constant. Under this condition we can expect no alteration in Planck's distribution [Lav91].

In any case, we must require the processes of photon creation and annihilation to be much more rapid than the expansion itself. This hardly seems possible in the inflation scenario. It is also worth observing that the walls can be totally absorbing, and hence totally emitting if isothermal conditions are complied with, or they can be perfectly reflecting, in which case a speck of carbon must be introduced into the cavity which plays the role of a black body [Pla14].

1.1.3 *Origin of CMBR*

It has been universally accepted by the big bang people that, provided the spectrum was not distorted by the addition of energy after the epoch corresponding to a redshift $z \sim 2 \times 10^6$, equivalent to a temperature of 5×10^6 K, the hot big bang produced an exact thermal spectrum [Par95]. Moreover this thermal distribution was to remain in tact even though the universe expanded by a factor of 10^{19}, and cooled to temperatures 2×10^6 lower than when the CMBR formed. From these assertions it is concluded that "Spectral measurements of the CMBR thus allow us to probe the physical conditions of the universe back to an epoch a few weeks or months after the big bang." [Par95]. It's nice to have such reassurances!

It is claimed that the close coupling of radiation and matter through electron-positron annihilation and re-creation together with Coulomb scattering surely kept the radiation thermal even when the universe was minutes old, corresponding to redshifts of $z \sim 10^9$. Thomson scattering, and its relativistic counterpart Compton scattering, preserve both the photon number and photon frequencies so that they cannot be implicated in the establishment of a thermal distribution of radiation.

In order to see why the events immediately following the big bang would be extremely hostile to the establishment of thermal equilibrium, it appears advantageous to review the conditions under which we might anticipate the formation of a thermal distribution of energy.

Let $K_\lambda d\lambda$ be the intensity of radiation in the wavelength interval from λ to $\lambda + d\lambda$. Furthermore, suppose that the energy flux reaches $d\sigma$, an element of the surface of the interior of a cavity — *not necessarily black* — whose walls are maintained at a constant temperature T. If $a_\lambda(T)$ is the fraction of energy absorbed by the walls in the interval from λ to $\lambda + d\lambda$, then the rate at which energy is absorbed is $\pi a_\lambda K_\lambda d\lambda$, where π is the sole remaining vestige of the integration over the solid angle $d\Omega$.

Now the rate of emission of energy in the same wavelength range into the cavity is $\pi e_\lambda(T)d\sigma$. Implicit in these expressions for absorption and re-emission is Kirchhoff's hypothesis that the rates depend on the temperature of the walls of the cavity and the wavelength, but are entirely independent of the material the cavity is made of. At thermal equilibrium, where the law of detailed balance holds, the two rates must balance one another,

$$\int_0^\infty a_\lambda K_\lambda d\lambda = \int_0^\infty e_\lambda d\lambda.$$

Studying cavities made of different materials, Kirchhoff came to the conclusion that detailed balancing should hold in *each* wavelength interval,

$$\frac{e_\lambda}{a_\lambda} = K_\lambda(T),$$

which depends only on the wavelength and temperature, independent of the nature of the cavity, whether it be its shape or composition. Cavities which are black have $a_\lambda = 1$, while cavities that are perfect reflectors have $a_\lambda = 0$. In the latter case, it is necessary to introduce a speck of charcoal, or some other black body that will absorb and re-emit at all wavelengths. This will allow an arbitrary distribution of energy to reach a thermal equilibrium distribution.

In the aftermath of the big bang there is no mechanism whereby a thermal distribution of radiation can be established. What about the effect of expansion on thermal radiation? In the last section we saw that the spectrum of black body radiation remains unaltered when the reflecting walls of the cavity are made to vary both adiabatically (in a mechanical sense) and reversibly. Although there are claims to the contrary, expansion of the universe, as envisioned by inflation, must alter Planck's law because it is adiabatic so the temperature falls as the radius of the universe grows. Thus, the inflationists are not dealing with an isothermal process where the walls are moved adiabatically (in the mechanical sense of slowness and not in the thermodynamic sense of heat impenetrability) and reversibly.

Looking to the heavens for examples of black body radiation is one of the oldest on the books. In the later half of the nineteenth century the American astronomer Langley attempted to determine K_λ by measuring the absorption and re-emission of solar radiation on a relatively cool surface of a planet. Qualitatively he found the same temperature dependent intensity maximum that he found with a copper radiator covered in lampblack. As a function of intensity, both curves tend to zero for both increasing and decreasing wavelengths. So here was the first example of black body radiation in the infrared for wavelengths of up to 5μ. No cavity was necessary, for interstellar space itself played the role of the cavity.

It has been argued, and almost universally accepted, that the further we go back in time, or the closer we come to the hot big bang, the more thermal the CMBR show should look. More precisely, there is a critical wavelength, λ_c, for which there is a transition from a Bose-Einstein distribution,

$$n_\nu = \frac{8\pi\nu^2/c^3}{e^{(h\nu+\mu)/kT} - 1}, \tag{1.2}$$

where μ is the chemical potential, for wavelengths $\lambda > \lambda_c$ to pure thermal radiation for $\lambda < \lambda_c$. The form given to the chemical potential is [IS75]

$$\mu = 1.4\frac{\Delta E}{E},$$

for small deposits of heat such that $\Delta E \ll E$, where $E = 4\sigma T^4/c$ is the energy density of black body radiation with σ as the Stefan-Boltzmann constant. Thus,

$$\mu = 1.85 \times 10^{14}\frac{\Delta E}{T^4}. \tag{1.3}$$

However the condition for chemical equilibrium is $\mu/T = $ const., or equivalently,

$$\frac{d\mu}{\mu} = \frac{dT}{T}. \tag{1.4}$$

From the Gibbs-Duhem relation,[1]

$$\frac{d\mu}{\mu} = \frac{dp}{\varepsilon + p}, \tag{1.5}$$

Stefan's law follows for the energy density, $\varepsilon \propto T^4$ when the thermal equation of state $p = \frac{1}{3}\varepsilon$ for the radiation pressure, and (1.4), are introduced.

---

[1] The Gibbs-Duhem relation follows from the desire to preserve the first order homogeneity of the entropy, $S = (E + pV)/T$, so that if $dS = (dE + pdV)/T$ then it must follow that

$$E\,d\frac{1}{T} + V\,d\frac{p}{T} = 0.$$

Introducing the energy density, $\varepsilon = E/V$, gives the Gibbs-Duhem relation (1.5) on the strength of (1.4).

This immediately contradicts any relation that would set the chemical potential as anything else than a linear function of the temperature. But let that not discourage us from finding the wavelength at which temperature 'step', ΔT_c, occurs in the transition from non-thermal to thermal radiation. Numerically it is found [CBZ91]:

$$\frac{\Delta T_c}{T} = 3.2\frac{\Delta E}{E}(\Omega_b h^2)^{-1/2} = 2.3\mu(\Omega_b h^2)^{-1/2},$$

and using the value of the density of baryonic matter, $\Omega_b h^2 = 0.02$, the thermal step should occur at $\lambda_c = 15$ cm, corresponding to 2×10^9 Hz which is far in the microwave region. This would be reasonable except for the fact that the chemical potential, (1.3), cannot be an inverse power of the temperature!

1.1.4 Black Body Fluctuations within the CMBR

Penzias and Wilson found in their original detection of CMBR that it was isotropic to within 10%. Within this region of anisotropies should lie information concerning the formation of large structures, or so we are told.

Reasoning based on the thermo-gravitational effect of Sachs-Wolfe led to the search of anisotropies, or 'fluctuations', as they are incorrectly referred to, set the scale at $0.1-1\%$. When temperature anisotropies were not found at this level, relative anisotropies in the temperature were pushed down to 10^{-4}. The introduction of non-baryonic matter, in order to achieve the critical density of a flat universe that was necessary for the inflationary scenario, led to a further reduction of the relative fluctuation, $\Delta T/T \simeq 10^{-5}$. Here nature seemed to come to rescue of the ailing theory, and COBE appeared to detect such anisotropies in the temperature range $\Delta T \simeq 30 \pm 5°\,\mu K$ over a large angular scale of $10°$.

According to Fred Hoyle and colleagues [FHN00], this would be the same as "if we were to explain minute fluctuations of the fire heating system of a boiler in terms of the shapes of the coals the stoker fed into it." Not quite, say the big bang people.

The model which emerges is again a black body, only this time a relativistic fluid comprised of photons that are hot enough to ionize hydrogen which form a single fluid that propagates at the velocity of sound, $c_s = c/\sqrt{3}$. According to the picture painted by the big-bangers, gravity attracts to compress the fluid into potential wells while the radiation pressure opposes this thereby setting up acoustic oscillations in the fluid. Regions that have reached maximum compression appear hotter and are supposedly visible as localized positive anisotropies in the CMBR. The modes arrive at the maxima or minima of the oscillation at recombination allowing the photons to decouple and travel from then on unobstructedly.

Unfortunately, these liberated photons have no longer any information to share on the nature of the CMBR. Neglecting the effect of gravity (which seems surprising because it is its effect that is trying to be modeled), the perturbed, linear equations of continuity and Euler are:

$$\frac{\partial \varepsilon}{\partial t} = -h_0 \nabla \cdot \mathbf{v}, \tag{1.6}$$

and

$$h_0 \frac{\partial \mathbf{v}}{\partial t} = -c^2 \nabla p, \tag{1.7}$$

respectively, where ε is the perturbed internal energy, $\mathbf{v}$ is the fluid velocity, h_0 is the unperturbed enthalpy density, and p is the perturbed radiation pressure.

Eliminating the fluid velocity between (1.6) and (1.7) results in

$$\frac{\partial^2 \varepsilon}{\partial t^2} = c^2 \nabla^2 p. \tag{1.8}$$

We can form the wave equation by relating the two perturbed quantities through $\varepsilon = (\partial \varepsilon/\partial p)_0 p$, where the subscript refers to the unperturbed adiabatic term. Or, we can express both terms in terms of the relative temperature anisotropy, $\Delta T/T \equiv \Theta$.

Since the unperturbed internal energy, $\varepsilon_0 \sim T^4$, and unperturbed pressure $p_0 \sim \frac{1}{3} T^4$, we have $\varepsilon/\varepsilon_0 = 4\Theta$, and $p/\varepsilon_0 = \frac{4}{3}\Theta$. The later will be appreciated as the Carnot-Clapeyron equation with latent heat $h_0 = \varepsilon_0 + p_0 = \frac{4}{3}\varepsilon_0$, and gives $(\partial \varepsilon/\partial p)_0 = \frac{1}{3}$ for an ultra-relativistic (photon) gas.

These substitutions enable (1.8) to be written as the wave equation,

$$\frac{\partial^2 \Theta}{\partial t^2} = c_s^2 \nabla^2 \Theta, \tag{1.9}$$

where the velocity of sound, c_s, is less than the velocity of light, c, by a factor $\sqrt{3}$ [LL59]. If one would like to express (1.9) in terms of Fourier components, Θ_k, because they are uncoupled, one could write

$$\ddot{\Theta}_k + c_s^2 k^2 \Theta_k = 0. \tag{1.10}$$

The solution to (1.10) is

$$\Theta_k(t) = \Theta_k(0) e^{k c_s t + \phi_k},$$

where $\Theta_k(0)$ is the initial amplitude, $c_s t$ is the distance traveled by the wave up to time t, and ϕ_k is the phase. On large scales, $k c_s t \ll 1$, and the perturbation becomes 'frozen,' while on small scales, $\Theta_k(t)$ displays oscillations in time.

We are then told that the relative temperature anisotropy can be developed in a series of associated Legendre functions, $Y_{\ell m}(\vartheta, \varphi)$,

$$\Theta = \sum_{\ell=0}^{\infty} \sum_{m=-\ell}^{\ell} a_{\ell m} Y_{\ell m}(\vartheta, \varphi),$$

where the $Y_{\ell m}(\vartheta, \varphi)$ are the associated Legendre functions. Someone appears to be missing the point here. The temperature anisotropies are supposedly the ones doing the scattering. What's being scattered are electromagnetic waves, the E and B fields. As a result of scattering, a plane wave is to be expanded in an infinite series of spherical harmonics, even though this may seem like forcing a square peg into a round hole.

The observed acoustic peak structure requires another element: the coherence of phases. This is supposedly provided by inflation where the modes outside the horizon are exponentially growing and decaying solutions. While these solutions cannot modify the causal dynamics that are taking place inside the horizon, the dominant exponential growing solutions provide for the temporal coherence of the modes by prescribing the initial conditions [AGP05].

With such a coherence, the modes arrive at the crests and troughs of the gravitational potential at recombination and are viewed as k-oscillations in the power spectra. The photons then decouple from the fluid and are free to propagate to us at the present time. However, it must be clear from the analysis that they will not retain any information regarding the black body spectrum because an integration over all frequencies has been made in order to obtain the global properties of internal energy and pressure.

So we have a picture of a single fluid with pressure supplied by the photons and inertia by the baryons. Moreover, said fluid is supportive of acoustic oscillations where both the internal energy density and velocity are oscillating functions of time. Now what?

We need polarization. Polarization of what? Polarization supposedly preserves the 'fossil' remnants about the expansion prior to recombination. As the temperature dropped below 10^9 K, an asymmetric expansion through Thomson scattering will, according to Rees [Ree68], produce a net linear polarization in the last few scatterings before recombination. Instead of declaring Rees's axially symmetric model a total failure both in practice and in theory, Nanos' [Nan79] null result was used to place an upper limit on the asymmetry of the Hubble expansion. This is typical of the big bang people who interpret null results as establishing upper bounds.

Thomson scattering cannot generate polarization after recombination because the universe has become transparent to photons which travel unhinderedly to the present epoch. Anisotropy appears as a sequence of hot and cold zones corresponding to maxima and minima, or light and dark zones, respectively, of the signal Θ_k. This is the rational behind expanding it in a series of associated Legendre polynomials, (1.14). The dipole term emits radiation perpendicular to the oscillations of the electrons. When integrated over all directions, the net polarization vanishes. Thus there is need of a sufficient quadrupole moment prior to last scattering in order to produce polarization. But what does the polarization of radiation have to do with the temperature anisotropy? And is the radiation basically that of a black body with an anisotropy superimposed?

1.1.5 *The Sachs-Wolfe Effect 'Scooped' by General Relativity*

As we have mentioned, in Penzias and Wilson's original discovery of the CMBR the anisotropy was set at 10%. If galaxy formation was to occur anisotropies had to be found. In the late 1960's, Silk and Sachs and Wolfe predicted anisotropies that were far too large in the relative 'temperature' fluctuation, $\Delta T/T$. Temperature anisotropies were later found that were a few parts in 10^{-5}, and these were implicated in structure formation. This is like reading the future in a Turkish coffee cup!

In 1967 Sachs and Wolfe showed that the standard model provided for large-scale fluctuations in the CMB temperature that were caused by a gravitational redshift of photons climbing out of a potential well. If the gravitational potential Φ varies from point to point so too will the temperature. A change in the potential $\Delta\Phi$ will produce (scalar) density fluctuations, and (tensor) gravity waves.

The radiation we observe coming from the CMBR underwent gravitational redshift, $\Delta\nu/\nu \sim \Phi$. And since $\Delta\nu/\nu = \Delta T/T$, the same can be expected for the CMB temperature [Pee93]. Notwithstanding the fact that the only relation between frequency and temperature is Wien's displacement law which applies to that frequency where the spectral density

curve of blackbody radiation peaks, such a relation between the temperature and gravitational potential already existed in general relativity in the
limit of weak Newtonian fields.

In a constant gravitational field a distinction must be made between the
conserved energy, E_0, and that measured by an observer which experiences
a static field, E. The two internal energies are related by [LL58]

$$E_0 = E\sqrt{g_{00}},$$

where the time component of the metric is $g_{00} = 1 + 2\Phi/c^2$ in the weak
gravitational limit.

According to the definition of temperature, the derivative of the entropy,
S, with respect to E_0 is the inverse of the temperature T_0, a constant
throughout. Alternatively the derivative of S with respect to E is

$$\frac{dS}{dE_0} = \frac{1}{T_0} = \frac{1}{\sqrt{g_{00}}}\frac{dS}{dE} = \frac{1}{\sqrt{g_{00}}}\frac{1}{T}, \qquad (1.11)$$

since the entropy cannot depend upon the presence, or absence, of a gravitational potential. Equation (1.11) provides a relation between the uniform
temperature T_0 and the varying temperature, T, depending on the strength
of the gravitational field, $T_0 = T\sqrt{1 + 2\Phi/c^2}$.

Only for weak fields do we have

$$\frac{\Delta T}{T_0} \equiv \frac{T - T_0}{T_0} = -\Phi/c^2 > 0, \qquad (1.12)$$

showing that the temperature is higher where the gravitational field is
stronger. This will work if E_0 is the internal energy measured at infinity,
while E is the internal energy measured anywhere outside a spherical shell
of mass M, which, according to Newton's shell theorem, is equal to the
mass as if it were concentrated at the center.

Moreover, according to the second part of Newton's shell theorem, the
energy E_0 could be measured anywhere inside the spherically symmetric
body since no net gravitational force is exerted by the shell on any object
inside the shell, regardless of the location of the object in the sphere.

The Sachs-Wolfe relation is even weaker than (1.12) since it deals with
a fluctuation in the gravitational potential, $\Delta\Phi$, caused by a fluctuation
in the mass, ΔM, which is indiscriminately related to a fluctuation in the

mass density, $\Delta\varrho$, viz.,

$$-\Delta\Phi = \frac{\Delta M}{M}\frac{4}{3}\pi r^2 = \frac{1}{2}(Hr)^2\frac{\Delta\varrho}{\varrho},$$

making $\Delta T/T \propto \varrho/\varrho$, where $H = \sqrt{8\pi G\varrho/3}$ is the Hubble constant.

Consequently, on large scales where the anisotropies are dominated by the gravitational potential, the Sachs-Wolfe effect is independent of the angular scale and is given by

$$\frac{\Delta T}{T_0} = -\frac{\Delta\Phi}{c^2} = \frac{1}{2}\left(\frac{Hr}{c}\right)^2\frac{\Delta\varrho}{\varrho}. \qquad (1.13)$$

Expression (1.13) does not need any special introduction since it follows directly from special relativity by perturbing the Kepler orbit, $v^2 = GM/r$.

Not so on smaller angular scales where it is supposed that a Doppler mechanism is responsible for the relative temperature fluctuation, $|\Delta T/T| = v/c$, where v is the velocity of scattered matter with respect to bulk matter on the surface of last scattering [Par95]. The velocity supposedly oscillates to give an acoustic wave with a wavelength of the order of the Jeans' wavelength. The acoustic peaks are expressed as terms of the higher harmonics of the spherical harmonics

$$T(\vartheta, \varphi) = \sum_{\ell,m} C_{\ell m} Y_{\ell m}(\vartheta, \varphi), \qquad (1.14)$$

where the polar coordinates ϑ and φ represent the right ascension and the declination, respectively, on the heavenly sphere. The $\ell = 0$ (scalar) mode is a density perturbation, $\ell = 1$ (vortex) describes the dipole anisotropy, while the $\ell = 2$ (tensor) is the next lowest term, the quadrupole.

Any observed anisotropy of the CMBR can theoretically be fitted by a sum of spherical harmonics [Par95]. However, in order for the temperature to be expandable in terms of associated Legendre polynomials, $Y_{\ell m}(\vartheta, \varphi)$, it must be a solution to Laplace's equation, at a constant radial distance, and enjoy the property of spherical symmetry. In fact, a gradient in the temperature should, and does, give rise to a heat flux. And if the divergence of the heat flux does not vanish, then the temperature should satisfy Poisson's equation, not Laplace's. Moreover there is absolutely no reason why a scalar temperature field should generate tensorial harmonics!

The fluid velocity has also been associated with the velocity of the local group moving with respect to the CMBR. From special relativity we know that all corrections to Newtonian mechanics comes in the ratio of

the Schwarzschild radius to the actual radius, and because it occurs in the
Lorentz contraction factor it must be proportional to at least the square of
the velocities. This effectively rules out linear terms.

1.1.6 *Inventing New Stokes Parameters*

According to Hu and White [WH97], and all those that have followed or
preceded in their footsteps, "Five quadrupole moments are represented by
$\ell = 2$, $m = 0, \pm 1, \pm 2$. The orthogonality of spherical harmonics guarantees that no other moments can generate polarization from Thomson
scattering."

Polarization of emitted radiation is a manifestation of the motion of a
charge, and Thomson scattering is scattering by a free charge. Unlike dipole
radiation where emission does not occur in the direction of oscillation, the
intensity of the radiation does not vanish anywhere in the observer's view
of the scattered radiation. Nevertheless the degree of polarization will vary
with the observer's viewing angle.

Dipole radiation of two opposite charges uniformly rotating at a fixed
distance from a common center results in circular polarization, something
that Thomson scattering can't achieve from free electrons. The velocity
of the earth's motion creates a Doppler shift of the radiation it receives
by shifting radiation to shorter wavelengths on one side, and longer wavelengths on the other, so as to increase and decrease the temperature, respectively. This has to be subtracted from the CMBR spectrum.

But the quadrupole exists and pertains to the CMBR spectrum. Instead
of there being opposite and equal charges rotating about a common center,
for quadrupole radiation, the charges are equal and have the same sign. To
achieve electroneutrality, two charges of the opposite sign can be placed at
the origin, but do not contribute to the radiation. Because of its circular
motion, each charge is accelerating toward the origin at all times. And
because the two electric fields created by each charge are equal in magnitude
but opposite in direction they will cancel out. The maximum radiation
occurs in the direction making an angle of $\pi/4$ with respect to the common
line connecting the two charges which are executing simple harmonic motion
in exactly opposite phases.

Although this radiation is linearly polarized it has nothing to do with
Thomson scattering of free electrons. In all expositions there is no rhyme
or reason why Thomson scattering off the 'last scattering surface', which
is a high brow way of calling the cosmic photosphere, be related to the

infinite number of multipole moments. There is also no reason, except from wishful thinking, why the quadrupole radiation should be associated with gravitational waves rather than oscillating charges about a common center.

Any unpolarized incident wave can be resolved into two incoherent linearly polarized components parallel and perpendicular to the scattering plane, $S_\ell = |S_\ell| e^{i\sigma_\ell}$ and $S_r = |S_r| e^{i\sigma_r}$, where the subscripts, as Chandrasekhar [Cha50] tells us, come from the last letters in the words parallel and perpendicular.

The four experimentally measurable Stokes parameters are:

- The total intensity
$$I = I_\ell + I_r,$$

- the difference between horizontal and vertical polarization,
$$Q = l_\ell - I_r,$$

- the difference between polarization at $\pm\pi/4$,
$$U = 2\sqrt{I_\ell I_r} \cos\delta,$$
and
- right- or left-handed circular polarization depending on whether
$$V = 2\sqrt{I_\ell I_r} \sin\delta,$$

is greater than or less than zero, where $I_i = |S_i|^2$, and $\delta = \sigma_\ell - \sigma_r$.

In Thomson scattering, the incident beam propagates along the z-axis and is scattered by a free electron. The Stokes vector of the scattered beam will be related to that of the incident beam by the Mueller matrix,

$$M = \frac{1}{2}\left(\frac{e^2}{mc^2R}\right)^2 \begin{pmatrix} 1+\cos^2\vartheta & \sin^2\vartheta & 0 & 0 \\ \sin^2\vartheta & 1+\cos^2\vartheta & 0 & 0 \\ 0 & 0 & 2\cos\vartheta & 0 \\ 0 & 0 & 0 & 2\cos\vartheta \end{pmatrix}. \quad (1.15)$$

If the incident radiation is I, and totally unpolarized, then the intensity of the scattered radiation, I' and linearly polarized component between $-\pi/2$ and 0 will be

$$I' = I\left(1+\cos^2\vartheta\right), \qquad \text{and} \qquad Q' = -Q\sin^2\vartheta, \quad (1.16)$$

where we have dropped the constant factor in (1.15).

The CMB fluctuation experts are under the impression that the Stokes parameters are able to distinguish among monopole, vector and tensor components of the temperature fluctuation. The temperature is expanded in a series of spherical harmonics, and the quadrupole component of the temperature pattern is taken as an example [HW97].

The quadrupole moment supposedly determines the polarization pattern through Thomson scattering. The necessity to consider multipoles arises in spin-1 particles, not for electrons which are spin-$\frac{1}{2}$ particles. Although the state $m = 0$ is represented as a spin vector in the plane normal to the z-axis, it is not possible to specify a direction in that plane since the spin vector precesses about the z-axis. This makes it necessary to consider quantities of higher rank than the polarization vector [KGBS91], which include a monopole, vector and a second-rank tensor. Since the monopole is specified through normalization, a complete specification of the polarization of a spin-1 particle requires eight *real* parameters.

According to the CMB experts, the temperature fluctuations have three geometrically distinct sources, all due to a photon-baryon fluid. This fluid causes scalar (compressional), vector (vortices) and tensor (gravitational) perturbations. These correspond to a spherical harmonic of rank $\ell = 2$ and orders $m = 0, \pm 1, \pm 2$, respectively. Nothing could be further from the truth!

The tensor components by themselves have five components, the vector another three, and finally a monopole which is taken care of by a normalization constant. That is, the monopole is not $m = 0$, the vector does not have components $m = \pm 1$, and $m = \pm 2$ are not to the components of a second-rank tensor!

The scalar perturbation of the temperature perturbation is described by an $\ell = 2$, $m = 0$ spherical harmonic $Y_2^0 \propto 3\cos^2 -1$, which supposedly describes hot photons flowing from hot regions into cold ones. According to Hu and White [HW97], "this pattern represents a pure Q-field on the sky whose amplitude varies in angle as an $\ell = 2$, $m = 0$ tensor or *spin-2* spherical harmonic"

$$Q = \sin^2 \vartheta, \quad U = 0. \tag{1.17}$$

This would correspond to linear horizontal polarization, in contrast to (1.16) which is vertically polarized but with the same intensity. The reference to a spin-2 particle, e.g., a graviton, is merely wishful thinking.

The vector perturbation is supposedly due to an incompressible, vortical flow which "produces a quadrupole pattern in the temperature with $m = \pm 1$, $Y_2^{\pm 1} \propto \sin \vartheta \cos \vartheta e^{\pm i\varphi}$."

What this spherical harmonic has to do with a vector is anyone's guess, but it has Stokes parameters,

$$Q = -\sin\vartheta\cos\vartheta e^{i\varphi}, \quad U = -i\sin\vartheta e^{i\varphi}. \tag{1.18}$$

Whereas (1.17) is a possible state of polarization, (1.18) is certainly not! First, the Stokes parameters must be real. Second, Q, the difference between vertical and horizontal polarization, must be a quadratic term of a circular function, just like the total intensity, I, which is the sum of the two. Third, as we have mentioned above, spherical symmetry allows one to set $\varphi = 0$, which would make U, in (1.18), a purely imaginary quantity! Now onto tensor perturbations.

It is absurd to claim that tensor fluctuations can be viewed as gravitational waves: "a quadrupolar 'stretching' of space in the plane of the perturbation... the accompanying stretching of the wavelength of photons produces a quadrupole temperature variation with an $m = \pm 2$ pattern $Y_2^{\pm 2} \propto \sin^2\vartheta e^{\pm 2i\varphi}$."

In addition, it is claimed that "Thomson scattering again produces a polarization pattern from the quadrupole anisotropy having Stokes parameters,"

$$Q = \left(1 - \cos^2\vartheta\right) e^{2i\varphi}, \quad U = -2i\cos\vartheta e^{2i\varphi}. \tag{1.19}$$

We are also asked to take "note that Q and U are present in nearly equal amounts for the tensors." Forgetting the φ dependency, this hardly seems to be the case since Q is real while U is imaginary. Moreover consulting (1.16), the expression for Q is what we would expect for the intensity I, and not a quantity which measures the difference between horizontal and vertical polarization.

The cases for which $U \neq 0$ are singled out, and are attributed the significance of the presence of gravitational waves. New modes, called 'electric', E, and 'magnetic', B, transform into one another as horizontal polarization becomes linear polarization at $+\pi/4$ while linear polarization at $-\pi/4$ becomes vertical polarization.

The analogy of the linear polarization coefficients, Q and U, with the E and B modes is not that all complete for the vanishing of the curl and gradient (divergence!) of the former and latter, respectively, are analogous to the vanishing of the second derivatives of Q and U. This is indeed surprising since the Stokes coefficients are functions of the spherical variables ϑ and φ, and in systems with spherical symmetry the latter can be set equal to zero.

In fact, the introduction of the E and B fields muddy the waters still further. E plays the role of some type of gradient, while B is supposed to mimic the curl operator. The Stokes parameters have been around since 1852, and if these operators, or whatever you want to call them, had any physics don't you think someone would have already introduced them?

In complete disregard of this, we are told that the Stokes parameters, Q and U "can be viewed as two shear components" [SZ98]. $Q > 0$ simply means that there is more horizontal than vertical polarization, while $U > 0$ indicates that the polarization is more akin to linear $+\pi/4$ than to $-\pi/4$. We are further told that E transforms as a scaler under parity, while B behaves as a pseudo-scalar. Furthermore, the B mode is the result of interchanging Q and U from their E modes definition.

Although the pair Q and U are not rotationally invariant, the pair E and B are. The hot and cold spots of the B field are supposedly where the vectors circulate in opposite directions. Since Thomson scattering deals with linear polarization, we must have $V = 0$ for both E and B fields. From the definition of V, this implies that the relative phase $\delta = n\pi$, for any integer n including 0. Moreover, for a B field we must have $Q = 0$, implying no preference for horizontal or vertical polarization, $I_\ell = I_r$. The sign of B is then determined by the sign of U, negative for odd n, and positive for even n including zero.

Now supposedly what is being scattered is the unpolarized radiation of the CMBR. The unpolarized incident light will result in linearly polarized light of intensity $I = \frac{1}{2}(I_\ell + I_r)$ with $Q = \frac{1}{2}(I_\ell - I_r)$, *and* $U = V = 0$. This rules out a B field for unpolarized incident radiation. It also rules out incident light that is polarized perpendicular to the scattering plane, and parallel to the scattering plane where $U = V = 0$, but $Q < 0$ in the former case while $Q > 0$ in the latter.

Moreover if the incident light is obliquely polarized with respect to the scattering plane at an angle $\pi/4$, then the scattered light will, in general, be elliptically polarized meaning that $V \neq 0$, so that according to the CMBR experts we are out of the realm of Thomson scattering.

The situation is even worse for an E mode whose sign is opposite to that of Q while requiring that $U = 0$. U vanishes in general for n half-integral, but such a condition would contradict the vanishing of V where n must be integral or zero. The only other way out is to require that the polarized light be completely horizontal in which case $Q > 0$ implying that $E < 0$, or completely vertical, $Q < 0$ and $E > 0$.

So, contrary to what has been claimed, polarization produced by Thomson scattering of radiation with non-zero quadrupole moment at last scattering will not tell us what part of the temperature anisotropies have arisen from vector and tensor perturbations.

1.1.7 Quadruple Polarization from Thomson Scattering?

Thomson scattering of photons by free electrons has supposedly taken place at the last scattering surface some 385,000 years after the big bang. At times small with respect to decoupling, the big bang scenario claims that the universe is hot enough to allow for the coexistence of free protons and electrons in a plasma soup. During this period the rate of photons to be scattered by free electrons is large with respect to the expansion rate of the universe. This permits photons and electrons to establish and stay in thermal equilibrium until the temperature falls to about 0.1 eV, occurring at a red shift of around $z = 1300$.

At this stage the protons and electrons combine into neutral hydrogen so that Thomson scattering ceases: the radiation is said to 'decouple'. The quadrupole is supposedly produced by free streaming photons at decoupling. Prior to decoupling, the quadrupole radiation is damped by the scattering of photons off free electrons leaving a monopole due to temperature fluctuations, and a dipole moment arising from a Doppler shift caused by the peculiar velocity of the free streaming photons. It is truly remarkable how much we know of what went on 385,000 years after it all began!

The total scattering cross-section, which is the ratio of the radiated intensity per unit solid angle, $d\sigma/d\Omega$, and the ratio of the incoming intensity divided by the cross-sectional area, I'/A,

$$\frac{d\sigma}{d\Omega} \bigg/ \frac{I'}{A} = \frac{3\sigma_T}{8\pi} \left| \hat{e}' \cdot \hat{e} \right|^2, \tag{1.20}$$

where the incident wave with linear polarization $\hat{e}'$ is scattered into a wave of linear polarization $\hat{e}$, and σ_T is the Thomson cross-section.

If the outgoing beam of light lies in the z-direction, we can choose the outgoing polarization vector components $\hat{e}_x$ and $\hat{e}_y$ to be perpendicular to, and in the scattering plane, respectively. Since the incoming wave is unpolarized, $I'_x = I'_y = I'/2$. The scattered wave will have Stokes parameters:

$$I = I_x + I_y = \frac{3\sigma_T}{8\pi A} I' \left(1 + \cos^2 \vartheta\right) \quad \text{and} \quad Q = I_x - I_y = \frac{3\sigma_T}{8\pi A} I' \sin^2 \vartheta, \tag{1.21}$$

where ϑ is the angle between the outgoing and incoming beams. The other two Stokes parameters are $U = 0$, and V is always zero since Thomson scattering cannot result in circular polarization.

Not asking why we should be interested in multipole moments in such a simple process as Thomson scattering, we are assured that the incident radiation can be expanded in spherical harmonics,

$$I'(\vartheta, \varphi) = \sum_{\ell m} a_{\ell m} Y_{\ell m}(\vartheta, \varphi),$$

where $a_{\ell m}$ are the expansion coefficients of the spherical harmonics, $Y_{\ell m}(\vartheta, \varphi)$. Each angular momentum level, ℓ, is $2\ell + 1$ degenerate with the magnetic quantum number, m, varying from $-\ell$ to $+\ell$.

The Stokes parameters of the outgoing radiation, when integrated over the solid angle, are found to be

$$I(\hat{z}) = \frac{3\sigma_T}{16\pi A}\left[\frac{8}{3}\sqrt{\pi}a_{00} + \frac{4}{3}\sqrt{\frac{\pi}{5}}a_{20}\right] \quad \text{and} \quad Q(\hat{z}) = \frac{3\sigma_T}{4\pi A}\sqrt{\frac{2\pi}{15}}a_{22},$$

since the coefficients $a_{\ell m}$ are real. If the outgoing radiation makes an angle β with respect to the z-axis, the incoming radiation beam must be expanded in a coordinate system rotated through an angle β. The relation between the rotated coefficient $\tilde{a}_{22}$ and the the original coefficient a_{20} will be

$$\tilde{a}_{22} = \frac{\sqrt{6}}{4}a_{20}\sin^2\vartheta.$$

The resulting Stokes parameter will then be

$$Q(\hat{z}) = \frac{3\sigma_T}{8\pi A}\sqrt{\frac{\pi}{5}}a_{20}\sin^2\beta. \tag{1.22}$$

We can now appreciate that (1.22) is identical to the original Stokes parameter in (1.21) when we identify the intensity of the incoming radiation as $I' = \sqrt{\pi/5}a_{20}$, and set the Euler angle $\beta = \vartheta$. This still does not answer the question of what is the need to consider unpolarized quadrupole radiation and for it to get Thomson scattered, unless it is to claim that this is the origin of gravitational waves, and to relate it to the polarization of the CMBR.

Now to answer the question of what has Thomson scattering to do with a quadrupole moment? Simply stated: absolutely nothing! This was a red-herring introduced by Rees [Ree68] in his ridiculous assertion that spontaneous and induced Thomson scattering can maintain a black body spectrum of the primeval radiation. On the contrary, Thomson scattering does not change the frequency of radiation. The amount of energy transferred from a photon to an electron is $\Delta E = (\hbar\omega/mc^2)\hbar\omega$. If the condition $\hbar\omega \ll mc^2$, the cross-section will no longer be Thomson, and we will be dealing with Compton (inelastic) scattering.

The total cross section for an oscillating charge at frequency ω_0 with damping η is

$$\sigma = \frac{8\pi}{3}\left(\frac{e^2}{mc^2}\right)^2 \frac{\omega^4}{(\omega^2 - \omega_0^2)^2 + \eta^2\omega^2}. \tag{1.23}$$

For frequencies $\omega \gg \omega_0$, the restoring force becomes completely irrelevant and the electron behaves as if it were free. The cross section (1.23) reduces to that of Thomson which shows a quadratic dependency on the frequency. This had already been derived by Barkla in his spectroscopic investigations.

In the opposite extreme where $\omega \ll \omega_0$, the cross-section (1.23) becomes proportional to the inverse fourth power of the wavelength. We then have Rayleigh scattering in which the dipole moment of the oscillating electron is described in terms of a static polarizability as a result of a static electric field.

Finally, when $\omega \sim \omega_0$, the cross section reduces to a Lorentzian form,

$$\sigma \simeq \frac{8\pi}{3}\left(\frac{e^2}{mc^2}\right)^2 \frac{\omega_0^2}{4(\omega - \omega_0)^2 + \eta^2},$$

which is the Fourier transform of exponential decay in time. In no case whatever does the cross section (1.23) give rise to a quadrupole moment, or implicate one!

1.2 Does the CMBR Keep Fossil Records?

The book of *Genesis* does a much better job of the creation of the world than the ingredients that are required for the formation of structure in the early universe. The density inhomogeneities can only begin to grow after the universe has made the transition from a radiation to a matter dominated universe. However baryons cannot be implicated in the growth of inhomogeneities until their 'decoupling' with photons which comes too late in the universe's evolution to provide a satisfactory explanation of galaxy formation that we observe today.

Thus it became necessary to use one's imagination and literally invent new matter in the form of non-interacting relic WIMPS, an acronym standing for weakly interacting massive particles. Because they are required to contribute to the density of the universe, and yet not be observed, WIMPS have been labeled as 'dark matter'. The list of suspect relics is long and impressive; it includes exotic particles like axions and neutralinos, light neutrinos, primordial black holes, super heavy monopoles, etc. They are further subdivided into the categories of 'hot' and 'cold' dark matter. These combine through adiabatic, isocurvature, or cosmic string perturbations to produce the structures we observe today. The nomenclature is impressive even if the physics isn't.

Support in favor of dark matter has come from the measurement of the rotational velocities of the arms of spiral galaxies which do not fall off with distance, and the gravitational lensing of light by matter which is apparently not present. The former is well accounted for by a modified form of Newtonian dynamics (MOND), as we shall see in §1.4, and there is nothing mysterious or strange about it. Rather, we shall appreciate that MOND is a misnomer since it does not modify the two-body interaction in Newton's law, but goes beyond it in taking multiple-body interactions into consideration [Lav95].

What the big bang people are faced with, in essence, is why the universe is so smooth, from the viewpoint of the CMBR, and yet so lumpy, from the perspective of structure formation. Rather than looking for observational evidence, the big-bangers, and more precisely, the inflationists, take the easy way out and define the type of process they want to deal with. The homogeneity problem presents itself already in the initial observations made by Penzias and Wilson. The noise they measured was uniform in all directions. Since the cosmic horizon was smaller in the past than it is now, in order to obtain such uniformity meant that signals had to propagate faster than the speed of light. Recalling that the radius of the horizon is the product of the speed of light times the time interval elapsed, this meant that perturbations had to cross the horizon and at a later date re-enter the horizon, which had time to increase.

Density perturbations outside the horizon cannot have any effect on the goings-on inside the horizon. For it would destroy causality. It is only when the perturbations re-enter the horizon that causality is restored. In order to separate points that were in causal contact, the 'space' underneath their feet has to expand faster than the speed of light. Now special relativity teaches us that nothing can travel faster than the speed of light. So how can space itself travel faster than the speed of light? The answer the inflationists give is that since space is not a body there is nothing restricting it from carrying any two points in the universe out of the horizon not by moving them *through* space, but, rather, *with* space.

But, is it space that we are talking about? the good-old fashioned three dimensional variety that we are familiar with? General relativity remains uncommittal over what the coordinates in its non-definite metrics stand for, but insinuates that the scale factor in the Robertson-Walker metric is the true radius of the universe. But who is to say that it is? And if it isn't then what is expanding exponentially? That's the beauty of general relativity: it can wiggle its way out of any uncomfortable situation simply through denial.

Uniformity is thereby 'solved' by postulating a period of rapid growth giving rise to faster than light expansion of 'space'. This irons out any non-Euclidean curvature that space could have had making it perfectly flat by bringing the density ever so close to its critical value. This is what is referred to as 'fine tuning', which without inflation would seem like a dark mystery. Even with inflation, it appears as an even a darker mystery as how can the act of expansion 'create' space out of nothing?

The inflationists talk of a critical 'density', while, at the same time, admitting that in Einstein's general relativity gravitational effects are not only caused by mass and energy but also by pressure. This is a far cry from his 'special' theory which claims that mass and energy, if not the change in energy, are equivalent. The pressure acting on the surface of a body of finite extent was brought in by Planck which converted mass–energy equivalence into mass–enthalpy equivalence.

With pressure as an energy density in the general theory, it can turn out to be negative which generates a gravitational 'repulsion' causing the universe to expand, rather than contract which is what common sense would tell us. Nevertheless, it is bad enough to confuse the rest energy density with the density of internal energy, but to append a negative contribution by the pressure when it comes to expansion, and not to take it into account in determining the critical density, is downright skulduggery.

One could argue that hyperbolic velocities are not constrained by the constancy of the speed of light, where they are related to the radius of curvature, becoming infinite as the speed of light tends to infinity, so that there is nothing illogical about such speeds. Yet, if the purpose of the rapid, faster-than-light expansion of space is to iron out curvature then such non-Euclidean velocities cannot be invoked. Nor do we know when 'enough is enough,' or if inflation might start anew at any time.

It all depends who is at the controls and why the explanation in *Genesis* is so much more appealing than inflation. At least there we *know* Who was in control, and Who was pushing the buttons. This is not to say that the Bible gives a scientific explanation of creation, far from it, but neither does the big bang scenario with inflation tagged on.

The big bang people still haven't answered the question of where the primordial seeds for structure formation came from? According to them, the primordial density perturbations could not travel faster than the speed of light, making the process of structure formation too slow, given the time since the big bang occurred. If the initial universe was of the size of atoms then the (probabilistic) laws of quantum mechanics should have been applicable. (The same appeal was made by Einstein when he realized the limitations of his general theory of relativity.)

In the defense of quantum mechanics are the uncertainty relations between conjugate variables, for if we know with unlimited precision the position of a particle we have to be totally ignorant about its momentum. Inflation takes these uncertainties and blows them up into large scale fluctuations whose size extends well beyond the minuscule horizon of the early universe. There they sit immobile waiting for the cosmic horizon to catch up and overtake them.

Specifically, the fluctuations need to be gaussian for which any statistical quantity can be specified in terms of a power spectrum. The only salvation of such an approach is that it can be refuted through observation. But this depends on who is doing the observation!

A worse choice couldn't have been made in implicating gaussian fluctuations in the process of structure formation. Gaussian fluctuations deal with small deviations from a time independent state for which the most probable value coincides with the average value [Lav85]. This is hardly what is needed to create structures from small deviations from deterministic behavior. Moreover gaussian fluctuations are determined by just two moments, which is quite exceptional for a probability distribution which usually requires an infinite number of moments in order to reconstruct the probability distribution.

An intuitive measure of the degree of inhomogeneity is afforded by the root mean square density fluctuation. It can be calculated in terms of the Fourier transform of the variance, known as the power spectrum. Because we are ignorant of the origins of the density fluctuations, it is usually assumed that the fluctuations do not have a preferential scale; that is, the power spectrum is given by a power law in the wave number k, i.e.,

k^n. The power n is chosen so as to render the density fluctuation constant as it crosses the horizon; this is referred to as the Harrison-Zel'dovich (HZ) spectrum.

It is truly amazing that while the HZ hypothesis preceded inflation by more than a decade it was found to be completely compatible with it. But don't be too surprised; inflation was found to be compatible with a lot that preceded it, like it was tailor-made to fit in.

The idea behind the HZ spectrum is that fluctuations arising in the primordial density had to have had the same probability no matter what their size. This indiscriminate policy gave equal weight to both large and small scale perturbations so that both small and large scale structures would have equal probability in forming. The same democratic policy was carried over in the inflationary scenario which makes the universe flat, by getting rid of 'cosmic' (geometric?) curvature, and made it uniform, while taking care to introduce small ripples in spacetime that could explain the emergence of large scale structures. And inflation saw to it that the democratic abiding fluctuations were well peppered throughout the cosmos.

Two types of fluctuations were envisioned: those in which radiation and matter are perturbed simultaneously, and those in which matter is perturbed alone leaving radiation, and hence, the temperature, constant. For obvious reasons the former are referred to as 'adiabatic' fluctuations, while the latter are known as 'isocurvature' fluctuations.

Adiabatic fluctuations leave the ratio of the baryon to photon number, n_b/n_γ, intact, but allow the density and pressure to vary. Since, according to Einstein's equations, the latter are source terms in those equations, they will modify the geometry, and therefore, affect the curvature. In contrast, isocurvature fluctuations — as the name implies — leave the curvature intact.

They do so by allowing a play-off to occur between density and radiation under the conditions that the total mass-energy, and curvature, remain constant. Since radiation is hardly perturbed, the temperature $T = $ const., and so these types of perturbations are also known as isothermal fluctuations. Clearly, inflation favors adiabatic fluctuations — at least during the period in which it is in operation.

The fractional perturbations in radiation and matter are related by

$$\left(\frac{\delta\varrho}{\varrho}\right)_r = \frac{4}{3}\left(\frac{\delta\varrho}{\varrho}\right)_m, \tag{1.24}$$

where the subscripts refer to radiation and matter. What is being confused here is the energy density of black body radiation which varies as T^4 and the photon number density which varies as T^3, which has nothing to do with the density of matter. This puts into serious doubt the label 'adiabatic perturbations' which should also be considered to be isothermal perturbations. But perturbations can't be adiabatic and isothermal at the same time unless one is considering absolute zero!

It is commonly assumed that if these adiabatic perturbations are present, the variations in the density and pressure will cause the curvature of spacetime to vary since density and pressure are the sources in the Einstein equations. Agreed, but Einstein's equations are adiabatic and hence cannot treat radiation, which needs isothermal conditions, so that (1.24) does not apply.

Statements like "photon fluctuations become a sound wave in the baryon-photon plasma," and "just before crossing [the horizon], the temperature fluctuation remains imprinted upon the photons giving rise to an anisotropy" [KT90] are meaningless strings of words that supposedly are meant to convey some important physical phenomenon. Fluctuations by their very definition don't leave fingerprints!

Adiabatic perturbations are 'honest-to-God' perturbations because $\delta\varrho \neq 0$, whereas isocurvature perturbations are not since $\delta\varrho = 0$ [KT90]. The adjective is totally meaningless. Adiabatic perturbations equate the relative fluctuation in the number density of any species of particle, n, with the entropy density of radiation, s, viz.,

$$\frac{\delta n}{n} = \frac{\delta s}{s}, \quad \text{which supposedly implies} \quad \frac{1}{3}\frac{\delta n}{n} = \frac{\delta T}{T},$$

asserting that "the magnitude of the temperature fluctuation is one-third that in any species" [KT90]. Nothing could be further from the truth. The only thing that can be equated with the entropy density of radiation is the density of photons, n_γ.

Kolb and Turner [KT90] argue, although unconvincingly, that since the number of some species, N, in a comoving volume is $N = na^3$, and the entropy density varies as $s \propto a^{-3}$, that N can be expressed as $N = n/s = \text{const}$. From this they conclude "the fluctuation in the local number density of any species relative to the entropy density ... vanishes; hence the name adiabatic." But, the entropy density is not proportional to the inverse volume so the conclusion that the number of a certain species moving in a comoving volume element remains constant is totally vacuous.

Rather, the condition for adiabaticity is $aT =$ const., so that the Kolb-Turner condition is not a condition for adiabaticity at all.

The relative entropy density fluctuation can be written as

$$\frac{\delta s}{s} = \frac{\delta(S/N_\gamma)}{S/N_\gamma} = \frac{\delta S}{S} - \frac{\delta N_\gamma}{N_\gamma} = 0, \qquad (1.25)$$

where $N_\gamma = N_\gamma(T)$ is the number of photons at temperature T. Since both the entropy, S, and the number of photons, N_γ, vary as T^3, the relative fluctuation in the entropy density, (1.25) vanishes. Hence, as (1.25) clearly shows, such perturbations are *not* adiabatic, but rather isothermal — a name reserved by the big-bangers for isocurvature perturbations ("terms which are routinely used interchangeably" [KT90]). To compare the entropy of radiation with any other quantity would be like comparing apples to bananas. If they can't even get their perturbations right how can they get their scenarios right?

Later on in their section on 'Foreplay,' Kolb and Turner [KT90] discuss isocurvature fluctuations which according to them, are not 'honest-to-God' fluctuations because $\delta\rho = 0$, and "are not characterized by a fluctuation in the local curvature." Now, $\delta(n/s) \neq 0$ is supposed to correspond to a fluctuation in an equation of state: "the local pressure depends not only on the density ρ, but also upon the composition."

The total energy density is incorrectly taken as the sum, $\rho_{tot} := \rho + \rho_R$, where $\rho = mn$, the mass density of any species, and $\rho_R \propto T^4$, is the energy density of black body radiation. What does the mass density have to do with an energy density? Nothing. If it were the rest mass density then why would the local pressure depend on it as they previously claimed? Although this is sloppy reasoning to say the least let's follow it through to find out where it leads them.

The reason that $\delta(\rho + \rho_R) = 0$ for super-horizon fluctuations is because "causality precludes the re-distribution of energy density on scales larger than the horizon." For it is only when the "the isocurvature mode becomes sub-horizon sized [that] fluctuations in the local pressure 'push' the energy density around and convert an isocurvature fluctuation into an energy density perturbation." If super-horizon fluctuations do not have any causal consequences how do we know they are there? And what is preventing them from re-entry even at some time in the future?

The condition that $\delta\rho_{tot} = 0$ implies

$$\frac{\delta T}{T} = -\frac{1}{4}\frac{\rho}{\rho_R}\frac{\delta n}{n} \qquad (1.26)$$

since the mass density $\rho = mn$, and the mass m is constant. Kolb and Turner make the absurd statement that "the size of the compensating temperature fluctuations is determined by ρ/ρ_R and $\delta n/n$." For it is abundantly clear that (1.26) is nonsense: The density, n, has to depend upon a power of the temperature less than 3, which is reserved for massless photons. This would make the mass in (1.26) temperature dependent and consequently $\delta m \neq 0$, contrary to assumption.

Next, Kolb and Turner ask us to recall that n/s 'measures' the number of species in a 'comoving volume element'. It is therefore 'natural' to define the relative fluctuation as $\delta \equiv \delta \ln(n/s)$. Then they conclude: "At early times when $\rho_R \gg \rho$, the compensating temperature fluctuation is small:

$$\frac{\delta T}{T} \simeq -\frac{1}{4}\frac{\rho}{\rho_R}\delta \ll \delta;$$

hence the name isothermal." But they just got through asking us to remember that the entropy density, $s \propto T^3$, and is far from being a constant as they now suppose in the above equation. It is therefore *not* true that the fluctuation in the local number density is just $\delta n/n \simeq \delta$.

'Early' times are now compared to 'late' times when "the universe becomes matter dominated [where] ρ becomes comparable to, and then greater than, ρ_R, and the temperature fluctuation becomes comparable to δ." We are left scratching our heads on how can a mass density be compared with radiation density; the same would go for a rest energy density.

The situation envisioned for isocurvature modes that enter the horizon after matter domination is "slightly different," but just as ludicrous. Just before horizon re-entry, $\delta T/T \simeq \delta/3$, whereas before it was the negative of this quantity, and the "temperature fluctuation remains imprinted upon the photons, giving rise to an anisotropy in the CMBR temperature." It is as if the photons wear T-shirts upon which imprints can be made by ink produced by temperature perturbations.

Once the perturbation is gobbled up by the horizon things get interesting — or so the inflationists would have us believe. Kolb and Turner claim that after horizon crossing, the difference between adiabatic and isocurvature becomes irrelevant (not that it was relevant to begin with). After border crossing, the electrons in the primordial plasma can become polarized. Because of the hot and cold temperature patches, the CMB can become polarized by electrons being hit by 'hot' and 'cold' radiation in orthogonal directions. The hotter the photon the more energetic it is. This is a whopper if ever there was one! The energy of a photon depends on its frequency, while the collective depends on its temperature.

The more energetic the photon, the higher its frequency is. A temperature can be related to a frequency at only one point: where the spectral density of black body radiation peaks. This is Wien's displacement law. So to even think that an electron being bombarded by hot and cold radiation in orthogonal directions will be polarized along the hotter radiation axis is an insult to anyone's intelligence.

It is even more ridiculous to associate this 'quadrupole' radiation with gravitational waves. The caveat is that this type of radiation can only come about at the time decoupling occurs where the photons are free to travel from denser to thinner regions of the plasma which the inflationists take as synonymous with hotter and colder regions. The mechanism of polarization is intricately related to the assumed oscillations of the primordial plasma. However these oscillations do not occur in space, but rather in *time*, so any reverse motion must be anti-causal. And how do you get standing waves by pinning the waves down at different points in *time*, rather than different points in space?

The spectrum of polarized radiation should also be related to the temperature fluctuation spectrum insofar as the peaks in one should correspond to troughs of the other. Supposedly, this would be a feather in the cap for inflation since the polarization spectrum would be different than that of the ordinary garden-variety scalar density perturbations. These are the so-called E and B waves that are rotated by $45°$ degrees. How electromagnetic radiation can result in gravitational radiation is not even broached, obviously so, for it would be impossible to explain. The same is true for the non-existent relation between thermal and polarized electromagnetic radiation, i.e., black body radiation is always unpolarized.

1.2.1 Why Should the CMBR be Blackbody Radiation?

It has been known since the time of Planck that black body radiation is a bounded thermal radiation field in a finite cavity. It is also known that the radiation is homogeneous, isotropic and unpolarized. So how does COBE measure a perfect blackbody spectrum at $2.728°K$ in unbounded space?

A perfect black body cavity is realistically unattainable. For it would consist of perfectly reflecting walls, maintained at a constant temperature, into which a piece of charcoal has been suspended. Planck's speck of charcoal is what thermalizes the radiation field and produces a dynamic equilibrium between the rates of absorption and emission of the radiation. Any time you see the principle of microscopic reversibility you know that the second law of thermodynamics is at work.

Had the reflecting walls been absorbent, they would have led to emission and absorption problems. Non-perfect walls have penetration depths for radiation, and, therefore, do not in general coincide with their geometric dimensions. A student of Lorentz established that the modes depend on the volume for only three geometrical shapes including those of a sphere and cylinder. However, without an enclosure there is no way of determining the number of modes in any given interval from k to $k + dk$, where k is the wave number, $k = \omega/c$.

Only if the number of modes in a volume V is given $D(k) = Vk^2/2\pi^2 dk$ will the Stefan-Boltzmann law

$$E_0(T) = \frac{\pi^2}{15} \frac{(kT)^4}{(\hbar c)^3} V, \qquad (1.27)$$

hold. As Planck rightly knew, (1.27) holds in the limit of high frequency and a large cavity volume such that $V^{1/3}\nu/c \to \infty$, or in the high temperature limit and a large cavity volume in such a way that $V^{1/3}kT/\hbar c \to \infty$. The former applies to the Wien limit, while the latter to the Rayleigh-Jeans limit, which holds if the linear dimension of the cavity, L, is much greater than ultra relativistic thermal wavelength, $\lambda_T = kT/\hbar c$.

When either of these two limits are not strictly satisfied there are corrections to the Stefan-Boltzmann law of the form [BH76]:

$$E(T) = a_0 V T^4 + a_2 V^{1/3}T^2 + a_3 T + a_4 V^{-1/3}, \qquad (1.28)$$

where the a_i are constants, a_2 and a_4 being shape dependent and $a_3 = k/2$, is shape independent, and there is no surface term, $E_1 \propto AT^3$, where A is the surface area in the case of a cavity with perfectly reflecting walls. The correction terms become important for radiation in the far infrared or sub-millimeter regions — an order below where the CMBR is! And if a characteristic length, L, of the cavity cannot be defined for a non-existent cavity, how then can the correct density of states be obtained in which modes would now be reduced by a linear term, L/c, in these regions? In other words, the total energy of radiation has to know how large the volume is.

The source of the CMBR is fitted by a Planck curve to better than 0.1%. However one may rightly ask why should a body radiating freely be the same as when it is enclosed in a black body cavity of perfectly reflecting walls [Wei60]. It was argued that emission is a property of matter alone so that the source field should not be altered, and still be described by Planck's function when the matter is extracted from a perfect enclosure without affecting its state since the only change is with regard to incident radiation.

This argument is rendered fallacious by the fact that Einstein showed that stimulated emission was necessary in order to derive the Planck curve from a dynamic balance between the rates of absorption and emission of radiation. Spontaneous emission alone resulted in Wien's distribution, and not Planck's. And since stimulated emission depends on the intensity of the incident radiation, it should be less in a freely radiating body than one in a black body cavity. This should put in jeopardy Kirchhoff's law since that law predicts a Planckian curve.

Kirchhoff's law claims that the ratio of the emissive to absorptive power is the same for all substances, being a universal function which can depend on the temperature as well as the frequency at which the equality is taken. It is known that it only applies to incandescent bodies and not luminous ones, which require an effective temperature different from the ambient temperature of the walls of the cavity. Hence it establishes a condition for monochromatic equilibrium where the intensity of light absorption is equal to that of emission, at each and every wavelength, where the latter is a function solely of the temperature and, of course, the wavelength of the radiation [Cot99].

Schwarzschild's equation for energy transfer is

$$\frac{1}{\kappa_\nu}\frac{dI_\nu}{ds} = I_\nu(s) - J_\nu(T), \qquad (1.29)$$

where I_ν is the intensity of light of frequency ν. The universal function J_ν was known to be a function of ν and T,

$$J_\nu(T) = \frac{\nu^3}{c^2}F\left(\frac{T}{\nu}\right), \qquad (1.30)$$

being based on strict thermodynamic arguments and Wein's displacement law, and the vertical distance $z = s/\cos\vartheta$, where ϑ is the angle that s makes with the vertical z. Actually, (1.29) fails to account for the fact that the temperature should also vary with height.

The "physical content [in (1.29)] is very slight" [GY89], but, whatever there is, it is buried in the absorptivity, κ_ν and the source function, $J_\nu(T)$. On the strength of 'local thermodynamic equilibrium' [Col77], the latter is identified as the Planck function. Then, and only then, will $I_\nu = J_\nu(T)$ be a statement of Kirchhoff's law. It occurs when the absorptivity $\kappa_\nu = 1$, which can be taken as the definition of a black body [Cot99]. We will now appreciate that it arises in a definite limit.

For a two level atom, like the one considered by Einstein, the absorptivity is

$$\kappa_\nu = \alpha \left(1 - \frac{n_2}{n_1} \right) = \alpha \left(1 - e^{-h\nu/kT} \right), \tag{1.31}$$

where the population of atoms in the upper level is $n_2 < n_1$, and the second equality assumes that the ratio of populations is given by the equilibrium ratio of their Boltzmann factors. The absorption constant, α, is independent of the height, z, for a homogeneous system. The exponential term in (1.31) is due to stimulated emission which Weinstein [Wei60] interpreted as negative absorption.

The solution to (1.29) is

$$\frac{I_\nu(z,\vartheta)}{J_\nu(T)} = 1 - e^{-\kappa_\nu z/\cos\vartheta}, \tag{1.32}$$

for $\cos\vartheta \geq 0$ in the upper half hemisphere. Weinstein [Wei60] refers incorrectly to the right-hand side of (1.32) as the 'emissivity', and claims that it is also equal to the absorptivity thereby vindicating Kirchhoff's law. His conclusion then follows that Kirchhoff's law is also valid for a freely radiating body.

However, the absorptivity will only be unity in the asymptotic limit of an infinite slab, as $z \to \infty$. This means that the photons will undergo multiple scattering, and many will not make it out alive.

This clearly shows that if the CMBR is black body radiation then the photons carry no information on how that state was reached. All information on how the thermal equilibrium state has been obtained has been obliterated by the condition that the optical path length be infinite. In this sense, the black body curve is a curve of maximum ignorance since it provides no information on how the system arrived in that state.

More generally for a gray body, whose walls are not self-reflecting, it is necessary to take into account the optical depth, $\tau = \alpha z$ for a homogenous material in which the absorption coefficient, α is depth independent.

The above analysis shows that only when the optical depth approaches infinity will the gray body become black.

It is also informative to consider the opposite limit of an optically thin medium, $\tau \ll 1$, where the mean free path of a photon is expected to exceed the depth of the gas layer. This can be thought of as being consistent with geometrical optics, for in this case (1.32) becomes

$$I_\nu(z, \vartheta) \simeq 2\frac{\tau}{\cos\vartheta}\frac{h\nu}{\lambda^2}e^{-h\nu/kT} = 2\frac{\tau}{\cos\vartheta\lambda^2}\bar{E}_\nu, \tag{1.33}$$

which defines the average energy,

$$\bar{E}_\nu = h\nu e^{-h\nu/kT}, \tag{1.34}$$

in the Wien limit.

We have chosen to express the density of states in (1.33) in terms of the wavelength, $\lambda = c/\nu$, so as to be able to introduce the absorption cross-section $\sigma_{abs}(\vartheta) = \lambda^2\cos\vartheta/4\pi$ in (1.33). Consequently, the energy absorbed by an antenna at frequency ν is

$$W_\nu = \sigma_{abs}I_\nu = \frac{\tau}{2\pi}h\nu e^{-h\nu/kT} =: \tau\bar{E}_\nu. \tag{1.35}$$

The Planck function has canceled the stimulated emission in (1.31) leaving spontaneous emission as the sole form of emission in (1.35). This occurs for optically thin systems, independent of the magnitude of the ratio, $h\nu/kT$.

Photons falling perpendicularly onto the antenna will be absorbed if they come within a radius c/ω. The area that will capture the photons will therefore be

$$\sigma_{abs}(0) = \pi\left(\lambda/2\pi\right)^2. \tag{1.36}$$

The maximum absorption occurs when the absorption length, $1/\alpha$, is comparable to the depth of the gas slab, z. Then integrating (1.35) over all frequencies relates the antenna temperature, T_A, to the energy absorbed by the heated resistor according to

$$W = \frac{kT_A}{2\pi}, \tag{1.37}$$

which is valid for each and every frequency interval throughout the whole of the radio range, extending from the longest waves to the far infrared region. In other words, the power spectrum is 'white', independent of the frequency. In fact, multiplying (1.37) by the band width, $\Delta\nu$, yields the well-known Nyquist's relation, $\dot{W} = kT_A\Delta\nu/2\pi$, relating the average power, $\dot{W}$, to the temperature of the antenna, T_A.

Nyquist's relation is the frequency, or ticker-tape, analog of temporal brownian motion with the frequency interval, $\Delta \nu$, replacing the time interval, Δt. More specifically, the mean square voltage is the frequency analogue of the temporal mean square displacement of a brownian particle, $\overline{(\Delta V)^2} = 4RkT\Delta \nu$, where R is the resistance that the brownian particle experiences. Since the coefficient of the bandwidth, $\Delta \nu$, is independent of the frequency, there is a direct proportionality between the mean square fluctuations in the voltage and the bandwidth.

This carries over to the average power which is then

$$\dot{W} = \overline{(\Delta V)^2}/4R = kT\Delta \nu.$$

Since the average power is independent of the resistance, R, there is a complete lack of knowledge of the way the system arrived in the equilibrium (fluctuating) state. In other words, the resistance R is related to the mechanism of transport (dissipation), and its absence is testimony to the ignorance on how the state was produced.

For each frequency interval, Nyquist generalized his relation to

$$W_\nu = \sigma_{\text{abs}} I_\nu = \frac{h\nu}{e^{h\nu/kT} - 1}, \tag{1.38}$$

which we have shown to correspond to infinite optical length. From (1.38) it is clear that Nyquist's theorem arises in the Rayleigh-Jeans limit, where $kT \gg h\nu$, and not in the Wien limit for which the inequality is inverted. Clearly (1.33) is the Wien limit which says that the *average* energy in each frequency interval is kT, and not the energies themselves. It is the Wien limit — and not the Rayleigh-Jeans limit — that constitutes the *classical* limit [Lav91].

This can be seen by solving (1.34) for the inverse temperature, employing the second law, $T^{-1} = dS/d\bar{E}_\nu$, and integrating to give

$$S(\bar{E}_\nu) = -k\frac{\bar{E}_\nu}{h\nu}\left(\ln \frac{\bar{E}_\nu}{h\nu} - 1\right), \tag{1.39}$$

which is the entropy of the Poisson distribution [Lav91]. In regard to the Kirchhoff function, (1.30), the average oscillator energy can be written as $\bar{E}_\nu = \nu F(\nu/T)$, where the function F turns out to be an exponential function of ν/T. Since the average energy is proportional to the heat absorbed by the antenna, the entropy will be a function solely of the ratio $x = \nu/T$, i.e., $S \sim xe^{-x}$. Moreover since the heat capacity, TdS/dT, must be positive, $\nu > kT/h$, again telling us that we are in the Wien region.

The absorptivity need not be limited to the exponential distribution, as in Einstein's case (1.31). If the absorptivity is due to thermal Doppler broadening then in place of (1.31) we have [GY89]

$$\kappa_\nu = \alpha \left(1 - e^{-Mc^2 \Delta\nu^2 / 2\nu_0^2 kT} \right), \tag{1.40}$$

where M is the mass of the particle, ν_0 is the central frequency and we have used the relation

$$\frac{\nu - \nu_0}{\nu_0} := \frac{\Delta\nu}{\nu_0} = \frac{u}{c},$$

to transform from the Maxwellian to the Doppler exponential.

Relation (1.40) requires that in place of the Kirchhoff function (1.30) there will now be the function

$$J_\nu(T) = F \left(\sqrt{\frac{Mc^2}{kT}} \frac{\Delta\nu}{\nu_0} \right), \tag{1.41}$$

not only highlighting the non-relativistic nature of the line shift, but moreover reflecting the non-universality of the function since, in addition to its dependence on frequency and temperature it also depends on the particular particle through its mass, M, and a particular spectral line, ν_0. There exists only the optically thick limit where the scattering mean free path is very short. This guarantees that the intensity is equal to the source function, (1.41),

$$I_\nu d\nu = F \left(\sqrt{\frac{Mc^2}{kT}} \frac{\Delta\nu}{\nu_0} \right) d\nu = \text{const.} \times e^{-Mc^2 \Delta\nu^2 / 2\nu_0^2 kT} d\nu, \tag{1.42}$$

even though the source function does not have the universal character prescribed by Kirchhoff.

Consequently, it is not necessary that the source function be a function only of the frequency and temperature in order that the shape of the absorption line be the same as the emission line if the incident light has a constant intensity in the region of the line breadth, which at half-maximum is

$$\delta = \nu_0 \sqrt{\frac{2kT}{Mc^2} \ln 2}.$$

The equivalence in (1.42) is between absorption and *spontaneous* emission when the line breadth δ is much greater than the natural line breadth, $2e^2 \nu_0^2 / 3Mc^3$. This requires the central wavelength, $\lambda_0 = c/\nu_0$, to be much greater than the geometric mean of the product of the electron thermal wavelength and the classical electron radius. The intensity distribution (1.42) decreases exponentially, exactly as the Wien distribution, (1.33), does. However, the latter can also occur in the optical thin limit whereas the former cannot because there is no associated Planckian function to Doppler broadening, which is entirely classical. In other words, there is no Planckian factor in (1.41) to cancel the exponential term in (1.40) which could be considered as stimulated emission in the optically thin limit.

1.2.2 *Ashes to Ashes, Dust to Dust*

In this section we argue that the Doppler and gravitational redshifts are accountable by the reddening of interstellar dust whose sub-micron particles are smaller than the wavelength of the radiation. We will also argue that the putative acoustic oscillations of the early universe, resulting in the periodic tug-of-war between gravitational attraction and repulsive radiation pressure, is none other than oscillations of the volumetric cross section of monochromatic light falling normally on a slab which can be looked upon as a sheet of dipoles.

At first, the cross-section increases with the thickness of the slab so long as all points on the slab are in phase. But once the thickness reaches a point where appreciable phase differences become possible, the cross-section will decrease with increasing thickness and oscillations will ensue. Superimposed on this series of broad and regularly spaced extinction maxima is a sharp and highly irregular structure, known as 'ripple' structure.

The ripple structure is also present in what is known as radiation-pressure-force spectroscopy. A particle is levitated by a radiation pressure counteracting the downward pull of gravity. It is moreover stabilized laterally by a laser beam, keeping the particle stationary at fixed height. If the wavelength is varied, the laser power for stable levitation will also vary, and so, too, will the radiation pressure. This gives rise to a complex ripple structure pattern.

This is not at all unlike the big bang scenario where, just after the big bang, the universe was filled with a dense and uniform plasma — one that would allow for the propagation of acoustic waves. The radiation pressure due to streaming photons acted to destroy density perturbations that were becoming denser and denser under the action of gravity. This push-pull type of behavior went on until the temperature fell to herald in the recombination era. Atoms then formed and matter became neutral. Radiation lost the power to halt gravitational collapse.

The universe then became transparent, matter started clumping together to form larger structures and photons were free to go their own way. Acoustic oscillations should then be ripples propagating in space like waves on a pond set in motion by a point source. But hold on, the contorted nature of the inflationary scenario is not that simple, and not at all straight-forward.

According to the inflationists, the CMB temperature fluctuations is a snapshot of the primordial plasma at recombination. Now what does temperature have to do with the acoustic oscillations resulting from the tug-of-war between gravity and radiation pressure? — which can certainly occur under isothermal conditions.

The answer is nothing, but temperature differences are supposedly what are measured on the celestial sphere. So suppose we have a measure of the CMB across the sky. At each place in the sky where there is a different temperature we can associate a different color so that the pattern of hot and cold regions will form a mosaic. Spots of the same size will then have the same wavelength. A plot of the average intensity versus their characteristic size might just give incoherent noise, or, if we are lucky, provide us with an idea of how intense the acoustic oscillations were in the early universe and provide a platform for inflation.

What is really happening is that we are putting the cart before the horse. There already exists a diagram of the intensity of thermal fluctuations versus the parameter they have been expanded into an infinite series of spherical harmonics, and the inflationists are trying to construct a story that supposedly describes it. But we immediately run into a snag when we try to push the musical string analogy further.

When we pluck a string not only will the fundamental be excited but also all those harmonics whose wavelengths are commensurate with the fixed length of the string. But where do we fasten the ends of the string to in the early universe? We don't: the constraints on the sound waves are not *spatial*, but *temporal* [Bal07]!

Acoustic oscillations have a limited duration, and end with recombination. Some waves will be caught in recombination in the phase of compression, others in the phase of rarefaction. Acoustic oscillations will alternate between peaks and troughs. Suppose a wave reaches its peak for the first time when it is caught by recombination. This perturbation will be the one with the largest possible wavelength, or the smallest possible frequency. This frequency can be identified as the fundamental, and all higher frequencies are harmonics. How we are to accomplish this in time is not specified — for the obvious reason that it can't. The perturbations that arrive in the compressional or rarefaction states leave a larger imprint on the CMB background, and therefore, give a larger contribution to the power spectrum.

Another basic difference between sound waves and acoustic oscillations of the primordial plasma is that the latter are not constrained to having harmonics of the fundamental frequency [Bal07]. However this will not detract from the fact that the waves corresponding to the fundamental frequencies will have the largest amplitudes since they are at the maxima of their oscillations at recombination. This is necessary in order to reproduce the qualitative feature of the principle peak even though it relinquishes explaining the successive peaks and troughs with ever diminishing sizes of the peaks.

Now we come to another caveat that hasn't been explicitly taken into account: In order that there be constructive interference, all perturbations of any given size had to start oscillating at the same time and with the same phase. This is a tall order that requires a director for the theatrics, and so enters inflation. Since perturbations are not limited by the speed of light they can easily surpass the horizon. Inflation is non-discriminatory and produces perturbations of every size. The smaller perturbations re-enter the horizon earlier than the larger ones where they are acted upon by gravity and the radiation pressure produced by streaming photons. Why they should want to return is another matter.

Since perturbations of the same size enter the horizon at the same time they will arrive in phase at recombination. Bingo, our problem of phase has been resolved, and not only that! It is because of inflation that we observe the regular features of the power spectrum so that these features are the fingerprints of inflation.

Finding the fingerprints of inflation gets a boost when the radiation bouncing off free electrons gets polarized. This requires the intensity of radiation to be different in different directions. Because of the temperature inhomogeneities the electrons hitting a photon in the NS direction may be hotter than those hitting the electron in the EW direction. If hot radiation means that it is more intense than cold radiation (why?) then the diffuse radiation will get polarized in the NS direction.

We are now told that polarization is intimately related to the acoustic oscillations of the primordial universe [Bal07], although sound waves can have only one possible polarization, namely the direction in which the wave is traveling (longitudinally). Disregarding this as a mere technicality, it is claimed that the acoustic peaks of the temperature fluctuations should correspond to troughs of the polarized spectrum. This would be additional evidence of inflation. But this is not all.

By merely observing temperature fluctuations it is not possible to distinguish between ordinary density perturbations and those created by gravity. However once polarization is taken into account things are different. Those that tend to compress or elongate should be related to ordinary density perturbations, while those that tend to rotate and twist should be related to gravity waves. The experts have called the former E waves, while the latter are known as B waves in obvious reference to electric and magnetic waves. But what is not so obvious is how electromagnetic polarization is transformed into gravitational polarization!

Just as surprising is the fact that the E and B modes are $\pi/4$ out of phase with one another. We would have expected a $\pi/2$ phase relation like s and p waves. The component of the electric field is parallel to the plane of incidence; the p-polarization being referred to as transverse-magnetic (TM), while the s-polarized light is said to be transverse-electric (TE).

It is not our intention to argue, like Hoyle and associates [FHN00], that carbon whiskers, if present in the gas clouds found within galaxies, are good candidates for the degradation of starlight. Rather, the anisotropies observed in the CMB can simply be thought of as particles which cause extinction of radiation.

On the strength of the optical theorem, extinction depends only on the scattering amplitude in the forward direction. The central peak in the plot of the extinction cross-section as a function of the size parameter, which is the ratio of the linear dimension of the particle to the wavelength, is due to the *interference* of the incident light and the forward scattered light [BH83]. Moreover there are a series of broad maxima and minima, called *interference structures* that oscillate, and superimposed on which there is a sharp, irregular structure called *ripple* structure.

The ripple structure is strongly damped when the absorption becomes large. The curve falling toward smaller wavelengths is a characteristic of absorbing particles which corresponds to the decay of the temperature power spectrum at high frequency that was predicted by Silk back in 1968. The greater the resolution, the greater the irregularities become. Ripple structure is a lot like brownian motion: dogs have fleas, fleas have fleas, *ad infinitum*.

The similarity between the temperature power spectrum as a function of the multipole moment, ℓ, and the volumetric cross-section as a function of the slab thickness, τ, is too similar to be passed up. Instead of representing acoustic oscillations of photons climbing out of potential wells and crawling back in we consider the interference of neighboring points on a slab representing a surface of dipoles.

In the beginning the volumetric reflection cross-section begins to increase monotonically with each new slab that is added. However there comes a point where there are appreciable differences in phase and the cross-section begins to decrease with slab thickness, reaching some minimum value where it begins to increase again. The cross-section begins to oscillate with ever decreasing amplitudes and decreasing distances between the peaks. *Since it is the radiation that is being observed, interference is a more reasonable explanation for the observed peak periodicity then some contrived, fictitious explanation of acoustic oscillations.* And because of interference, the total effect is not the sum of its causes.

Instead of Doppler and gravitational redshifts, we can consider the reddening of starlight caused by interstellar dust. The dust particles between stars extinguish blue light more effectively than red light resulting in a reddening of starlight that does not require relative motion nor a gravitational field. Reddening occurs when the size of the dust particles are smaller than the wavelength of radiation. If reddening does occur by Rayleigh scattering then the opposite effect of bluing should also be possible for a narrow range of particle sizes. Bluing would vanish as the dispersion in the particle radii increases.

In their frenzy to patch up the big bang scenario, the big bang people have turned the old adage that "a picture is worth a thousand words" on its head to mean that a thousand words must given to explain a picture so that it does not contradict a preconceived theory, no matter how outlandish it may be. Theories come and go, but causes remain!

1.3   Tally-ho, Inflation to the Rescue!

Hindsight, which is always twenty-twenty, allows us to appreciate that the unlimited speed by which neighboring points in space were able to distance themselves during the period of inflation actually takes its cue from the hyperbolic velocity law introduced by Hubble to measure the recession of galaxies.

When inflation has done its work (we know, but how does it know?), the early universe has blown up like a big balloon, and things have been homogenized. How does our universe get out of such a run-down, lethargic state?

The answer: A (first order) phase transition in which there is "a release of latent heat" that can warm the universe to where it was so that it can proceed with its nucleosynthesis. To make sure that the heat is sufficient, it is added that the universe has undergone a supercooling so that that temperature at which the phase transition takes place is far inferior to the temperature it would have normally taken place. And just think, all this is going on in a system undergoing adiabatic expansion!

A problem with this scenario (and there are others!) is that latent heat does not heat. In the words of Bohren [BA98] "To accept an explanation of temperature rise because of net condensation as a result of the release of some ethereal fluid is like eating sawdust instead of a proper meal: The sawdust fills your stomach but has no nutritional value."

1.3.1 *Hubble's Law and Superluminal Velocities*

The redshift is defined as

$$z := \frac{\lambda_o - \lambda_e}{\lambda_e} \equiv \frac{\Delta\lambda}{\lambda}, \tag{1.43}$$

where λ_e is the emitted wavelength and λ_o is the observed wavelength. By setting $z = v/c$, the relative velocity, one is implying that the classical Doppler effect is the culprit behind the change in wavelength.

If we combine this with Hubble's law,

$$v = H \cdot d, \tag{1.44}$$

where Hubble's 'constant', H, is the proportionality factor between the recession velocity, v, and distance, d. Introducing c times (1.43) for the recessional velocity in (1.44) gives

$$d = \int_0^z \frac{c \, dz'}{H(z')}, \tag{1.45}$$

where leeway has been given for values of the Hubble 'constant' that depend upon the redshift. Hubble's constant is hardly a constant.

According to Peacock [Pea99], the interpretation of (1.45) is "quite clear" for $v \ll c$. "What is not so clear is what to do when the Doppler shift becomes large. A common but incorrect approach is to use the special-relativistic Doppler formula and write

$$1 + z = \sqrt{\frac{1 + v/c}{1 - v/c}}.$$

This... is wrong in general."

Peacock attributes the error to a non vanishing matter density which causes gravitational redshifts so that he would write

$$1 + z_o = (1 + z_D)(1 + z_{gr}) \cdots,$$

where z_o is the observed redshift due to z_D, Doppler, z_{gr}, gravitational, etc. redshifts. This is only a pretext to come out and blatantly affirm that "$1+z$ tells us nothing more than how much of the universe has expanded since the emission of photons that we now perceive." Rather, this is nothing more than (1.43) with the scale factor $a(t)$ replacing the wavelength, and with $z = v/c$ we are back to our old expression (1.45) as a measure of distance.

The reference metric that is used is the Robertson-Walker metric,

$$ds^2 = -c^2 dt^2 + a^2(t) d\chi^2(z), \qquad (1.46)$$

where $\chi = \tanh^{-1} v/c$, and not the inverse hyperbolic sine as is almost universally employed, and erroneously believed to be correct. This is because the spatial metric must coincide with a known metric of constant curvature. The curvature is negative, as in Milne's universe, and the spatial metric coincides with the Beltrami metric [Lav11].

Now the radial distance along a time slice $(dt = 0)$ is $D = a\chi$. Differentiating this proper distance gives [DL01]:

$$v_{tot} = \dot{a}\chi + a\dot{\chi} = v_H + v_{pec}, \qquad (1.47)$$

where the total velocity, v_{tot}, has been decomposed into a recessional velocity, v_H, and a peculiar velocity, v_{pec}. The recessional velocity is given by

$$v_H = \dot{a}(t)\chi(z) = \dot{a}(t) \int_0^z \frac{c \, dz'}{H(z')}. \qquad (1.48)$$

Simplifying (1.48), and introducing the definition of χ result in

$$\chi = v_H/\dot{a} = c \int_0^z \frac{dz'}{H(z')} = \tanh^{-1} v/c. \qquad (1.49)$$

This clearly shows that whereas v is the Euclidean measure of the velocity, v_H is the corresponding hyperbolic measure.

Twisted and confusing statements abound in the literature. For instance, "the relative velocity cannot exceed c is only defined for particles whose separation is much less than cz/H. Any definition of relative velocities at distances larger than the curvature scale, where the Hubble law predicts velocities which exceed c, cannot have an invariant meaning" [Muk05].

This is what supposedly clarifies the 'superluminal' expansion in inflation. It is ironic since inflation purports to show that exponential expansion makes space flat, where the density is equal to the critical density, while it unintentionally uses velocities that belong to hyperbolic space with negative constant curvature.

Unlike the Euclidean relative velocity v/c, the relative hyperbolic velocity, χ, is not limited to values less than unity. In terms of a Hubble sphere, those inside recede at velocities less than the speed of light while those outside recede faster than the speed of light. According to inflation, the galaxies remain stationary, but it is the intergalactic space that they are found in which expands [Har00]. Supposedly this circumvents the thorny question that no object can travel faster than the speed of light — and space is no object. If the reader finds this hard to swallow he should not feel alone. However it serves as an example to what extent the big bang people are willing to go to save a theory that would be better off dying with some dignity.

But it does not stop here. Because of the expansion of space, an observer's light cone in space-time does not stretch out straight, as it would in a static universe, but contracts back to the big bang. "The observer, by looking in any direction, looks back into the big bang, and the light that the observer receives from the big bang is the cosmic background radiation" [Har00]. However for COBE to measure a black body spectrum, the CMBR cannot be a fossilized remnant of the past glow of the big bang, but rather an on going dynamic equilibrium between the absorption and emission of thermal radiation.

It is then claimed that the light the observer sees is continually and increasingly redshifted as the observer looks further back toward the big bang. But, whereas $z = v/c$ can only be used for $z \ll 1$, Hubble's law, (1.44), is valid without exception. This is undoubtedly a consequence of the fact that $z = v/c$ corresponds to the classical redshift, which at some point must be superseded by its relativistic generalization.

1.3.2   *Potpourri of Phase Transitions in the Early Universe*

The idea that phase transitions had something to do with changes in the early universe goes back to the Grand Unified theories (GUTS) of the seventies. Coleman proposed that the transition from a state dominated by the hyperweak force to a state of lower temperature governed by the strong and electroweak forces, consisting of a quark and lepton soup, occurred as a result of a phase transition. In this section, we don't contest so much the idea of a phase transition, but rather, what type of transition it should be.

The analogy drawn was to a transition between water and ice as the temperature is lowered. If it so happens that water is supercooled in such a way that it remains liquid below its freezing point, it can transform into ice spontaneously at a certain point with the possibility of liberating latent heat which can subsequently be used to reheat the system to a much higher temperature.

In terms of quarks and leptons this would mean that they fail to appear at the critical temperature of 10^{28} K, but the so-called 'elm', as Gamow called it, super cools into what Coleman referred to as a 'false' vacuum. For, according to Coleman, the false vacuum corresponds to a superheated fluid while the 'true' vacuum is analogous to the vapor phase.

Thermodynamic fluctuations are replaced by quantum fluctuations, and the decay of the false vacuum occurs with the formation of bubbles, which, if large enough, will continue to grow. "Once this happens the bubble spreads throughout the universe, like a cancer of space, converting the false vacuum to true," in Coleman's [Col77] own words.

The only problem with Coleman's proposal is that the transition from the false vacuum to the true vacuum is a second-order phase transition, and superheating and supercooling cannot occur in phase transitions higher than first. Second-order transitions are 'either or' transitions, while first-order transitions envision the coexistence of two phases simultaneously.

Coleman's actual proposal was to consider that an immense quantity of latent heat is released when the universe finally undergoes the transition from the false to the true vacuum thereby restoring the temperature to its temperature near to its original value, prior to the onset of inflation, but with the addition of quarks and leptons.

Guth borrowed the same idea by considering the false vacuum to be that state of the universe in which negative pressure causes the universe to accelerate. The association goes all the way back to de Sitter who found in

1917 the same effect to occur when the cosmological constant is appended onto Einstein's equation. The cosmological constant plays the role of an unphysical, constant pressure which is negative.

But there was no way de Sitter could end the expansion, which, because of its adiabatic nature, would have been accompanied by a sharp fall in the temperature. Guth used Coleman's idea that the release of latent heat when the universe undergoes a transition from the false to the true vacuum would warm the universe back up to temperatures which would allow nucleosynthesis to proceed as in the standard model.

If the volume and temperature are chosen as independent variables, the variation of the heat with respect to volume, at constant temperature, defines the latent heat, while its variation with respect to the temperature, at constant volume, is the heat capacity. The former is the latent heat while the latter is the sensible heat. These are two distinct, and different, forms of heat.

It is a common mistake in meteorology to assume that when water vapor condenses to form clouds latent heat is released into the atmosphere that warms the surrounding air around a new cloud droplet that causes instability resulting in thunderstorms.

Latent heat is available to perform work, but it does not cause a rise in the temperature. The heat must first be converted into sensible heat which can then cause a change in temperature but with no accompanying phase change. In other words, latent heat does not heat, and even worse, there is no latent heat in a second-order phase transition.

The hot big bang was rather successful in predicting the age of the universe as the inverse of the Hubble constant, and the abundance of ^{4}He. Less successful was its explanation (or lack thereof) of why the curvature of the universe is so small, or why is the density so close to the critical density if one believes in the Friedmann equation

$$\frac{k}{\dot{a}^2} = \Omega - 1, \tag{1.50}$$

where k is the normalized curvature, either $0, \pm 1$, $\Omega = \varrho/\varrho_c$ and the critical density is $\varrho_c = 3H^2/8\pi G$.

The standard model can also not predict why the ratio of baryons to photons is $n_b/n_\gamma \sim 10^{-9}$. Moreover how could non-causally connected regions at early times, $z \simeq 1000$, have the same density and temperature that they now do? How do you explain the isotropy of the CMBR over scales greater than the horizon at redshifts of the order of 10^3? And, finally, how do you introduce inhomogeneities into such a homogeneous structure that would be the seeds of galaxies? Easy, you resort to science fiction to bridge an otherwise unbridgeable gap.

Inflationary fiction attempted to do this by allowing the baryon number to go unconserved and prophesized that a period of exponential expansion took place shortly (very shortly!) after the hot big bang.

Taking the call from the Higgs field, 'new' inflation said let there be a scalar field, ϕ, whose nature is of no concern to us, equipped with a potential $V(\phi)$, which starts off at some non zero value, $V(0) \neq 0$ and 'slowly rolls' into a minimum at some non-zero value, ϕ_s.

The initial value of the potential, $V(0) \neq 0$, is thought to represent the energy density of the 'false' vacuum. It replaces the density in the Friedmann equation,

$$\dot{a}^2 = \frac{8\pi G}{3} V(\phi \simeq 0) a^2,$$

leading to an exponential solution,

$$a(t) = e^{\sqrt{\Lambda/3} t}, \tag{1.51}$$

just as in the case of the cosmological constant in the de Sitter case. All that was needed was to replace the cosmological constant, Λ, by $V(\phi \simeq 0)$. But who says that Einstein's equations are valid some 10^{-34} s after the big bang when they are not valid in much larger regions inside the Schwarzschild radius?

The universe did a 'slow roll' from an unstable equilibrium at $V(0)$ to a stable equilibrium at ϕ_s. The slow roll between $0 \leq \phi \leq \phi_s$ was required to allow inflation do its job. In other words, the time it takes to roll down is the period of inflation. The density, ϱ_ϕ, and pressure, p_ϕ, fields can be expressed in terms of the field kinetic energy, $\dot{\phi}^2/2$, and potential, V, as [Pee80]:

$$\varrho_\phi = \dot{\phi}^2/2 + V, \quad \text{and} \quad p_\phi = \dot{\phi}^2/2 - V.$$

A simple inflationary scenario assumes $\dot{\phi}^2 \ll V$. The field energy density and pressure are

$$\varrho_\phi = -p_\phi = V, \tag{1.52}$$

which is approximately constant at the start of the slow roll. From Einstein's equations we find

$$\frac{8\pi G}{3} \varrho = \frac{\dot{a}^2 + k}{a^2} = \text{const.},$$

and

$$8\pi Gp = -\frac{2a\ddot{a} + \dot{a}^2 + k}{a^2} = \text{const.}$$

Taking the time derivative of the first equation gives

$$\ddot{a} = \frac{8\pi G}{3}\varrho a,$$

which can simply be integrated to give

$$a(t) = \exp\left(\sqrt{8\pi GV(\phi \simeq 0)/3}\, t\right),$$

since $\varrho \simeq V(\phi \simeq 0) = \text{const.}$

Moreover the conservation of energy,

$$\frac{d}{dt}\varrho a^3 = -p\frac{da^3}{dt},$$

requires $\varrho = \text{const}$ so long as the equation of state (1.52) is in effect. But once in place what is there to change the equation of state (1.52)? We are therefore resigned to a state of *eternal* inflation!

Even if we were to accept this who is the master-mind behind it all? Natural science started with an effort to understand what physical phenomena is; in the twentieth century that aim was reduced to describing physical phenomena. It appears that in the twenty-first century we are willing to accept anything just so it gives what we presently believe is the correct description no matter how far-fetched that may be.

GUTS teaches us that all the forces including the strong and electro weak forces were unified at energies greater than 10^{14} GeV. This means that the big bang must have begun at thermal energies superior to this energy, and as adiabatic expansion unfurled, the universe cooled down allowing for a de-coupling of the forces. If the coefficient in front of time in (1.51) is replaced by E^2, where $E = 10^{14}$ GeV, then the universe will increase by some 100 e-fold in a time of $t = 10^{-32}$ s. Since the expansion is adiabatic, $T \propto a^{-1}$, the universe will have cooled to a lethargic state with $T \approx 0$. And since the universe is adiabatically isolated there is nothing to bring it out of such a state.

In Guth's original version, inflation ends with the onset of a first order phase transition that creates the entropy we see in the CMB. Linde and Albrecht and Steinhart subsequently modified this picture so that the transition from the inflationary regime to the classic Friedmann model was rapid enough to create the necessary entropy. Necessary for what?

At the end of inflation the baryon density is negligible for all intent purposes. In order to replenish the stock, the entropy produced at the end of inflation, by converting the latent heat of the phase transition into heat that would reheat the universe, would be converted into baryons. How latent heat heats, and how entropy is converted into baryons is left for the reader, or a magician, to decipher. Would it be too embarrassing to ask who exchanged the adiabatic walls surrounding the universe for diathermal ones?

Ah, but we forgot to include quantum fluctuations that make the field oscillate about the state ϕ_s. So a 'slow roll' toward that state has turned into a 'fast rock' about that state! Again it would not be inappropriate to ask who is at the helm? Now these oscillations have a job to do, and that job is to decay into particles, including photons. This supposedly causes a reheating of the universe to get it back on track with our present-day understanding of it.

The CMBR didn't originate in the big bang but was generated 10^{-32} s after it. Surely a lot happened in a very short time! It's not everyday we get to witness such events, and the question is why not? What was so special about that instant in time? Avoiding to answer this, but invoking the same quantum fluctuations, the big bang people insist that they are the origin of all the large scale structure we see today. I don't see how this is much better, or more satisfying than invoking a master designer.

1.3.3 *Latent Heat Doesn't Heat*

Undoubtedly the most precarious element in the inflationary scenario is how the universe gets out of its sluggish, run-down state at the end of inflation. Inflationists argue, from the time of Guth's first publication, that an invigorating first order phase transition is required to get the universe back on its feet and on the normal tract to its present day location. In the words of Guth, who undoubtedly took them from Coleman, [Gut81],

> Suppose the equation of state for matter ... exhibits a first-order phase transition at some critical temperature T_c. Then as the universe cools through the temperature T_c, one would expect bubbles of the low-temperature phase to nucleate and grow. However, suppose the nucleation rate for this phase transition is rather slow. The universe will continue to cool as it expands, and will then supercool in the high-temperature phase. Suppose that this supercooling continues down to some temperature T_s, many of orders of magnitude below T_c. When the phase transition finally takes place at temperature T_s, the latent heat is released. However, this latent heat is characteristic of the energy scale

T_c, which is huge relative to T_s. The universe is then reheated to some temperature T_r which is comparable to T_c. The entropy density is then increased by a factor of roughly $(T_c/T_s)^3 \ldots$, while the value of R [the radius of the universe] remains unchanged.

Although the idea was in the air for some time, Guth's paper is heralded as the 'shot heard round the world.' Rather, it seems more like a whimper, and in no time of the history of scientific publishing has so much unsupported speculation has been published in the literature, and in such a supposedly prestigious journal as *Physical Review D*. It appears as a last ditch attempt to salvage the unsalvageable. The nonacceptance of one 'suppose' and the entire scenario falls apart. It even falls apart without this.

Einstein's equations, upon which Guth based his analysis, are adiabatic. Adiabaticity provides a constraint on the variation of the radius of the universe with temperature that is violated in Guth's imaginary scenario. Being adiabatic, where does the universe find the enormous amount of entropy to increase?

Moreover, as we have repeatedly emphasized, latent heat can't heat. Black, who coined the word latent, wrote in his *Lectures on the Elements of Chemistry* published in 1803, some four years after his death and nearly forty years after he had written them:

> When we deprive a body of its fluidity ... a very great amount of heat comes out of it, while it is assuming a solid form, the loss of which heat is not perceived by the common manner of using a thermometer ... The extrication and emergence of latent heat, if I may be allowed to use these terms, appears to be concealed within the water.

Black is making the distinction to what is latent, or insensible to the touch, and sensible heat, which can be used to raise the temperature. Although this distinction already appears in caloric theory, it underlines the fact that latent heat arises from a change in the heat when the volume — and not the temperature — is varied [Lav09].

Even if we were to overlook all this, there would still be the problem that phase transitions do not increase the entropy of the two phase system. Entropy increases in *irreversible* — not reversible processes. Moreover, Guth has confused microscopic nucleation processes with macroscopic thermodynamics which is a 'black box', unable to discriminate the underlying processes that have led to the establishment of a phase equilibrium. Kinetics is a branch of science distinct from thermodynamics; the growth of bubbles belongs to the former, while the latent heat required to transform one mole of substance into a different phase lies in the realm of the latter.

Any discussion of the 'release of latent heat' necessarily requires the introduction of non equilibrium concepts like a temperature gradient. According to Fourier's law, which he published in 1822 in his *Analytic Theory of Heat*, the flow of heat is proportional to the negative of the gradient in temperature.

Consider an air-water interface at 0°C. Water existing as a liquid below 0° C is referred to as supercooled, but once it does begin to freeze ice will begin to form at the interface with the air, both of which are at the same temperature.

Water must give up its (latent) heat on freezing. On freezing water expands thereby increasing its volume. If the ice-water interface is at 0 degrees and the air-ice interface is lower because of supercooling then a thermal gradient will be formed across the ice. Heat will be conducted by Fourier's law across the ice, the temperature gradient will decrease because the slab of ice is increasing thereby expanding its volume. Since the overall energy is conserved less heat will flow across the interface. And the temperature at which it flows across the interface is the interfacial temperature which is necessarily lower than the freezing temperature of water for, otherwise, there would be no thermal gradient to begin with. So the latent heat cannot be at a much higher temperature than the normal freezing point of water. Latent heat must first be converted into sensible heat before heating can begin.

All this proves that the conversion of latent into sensible heat is not an efficient way of heating. Moreover any increase in entropy will occur through the irreversible process of heat transport which is way beyond the realm of an equilibrium phase transition.

Thus, it is safe to conclude that Guth's idea of reheating, and re-entropizing the universe, is still-born. And since an equilibrium transition is being contemplated with a pre-existing metastable phase there is no way of assigning a temporal progression to such processes. For all we know, we could still be in the reheating stages, or if not, inflation could be eternal!

1.4 Is MOND really Modified Newtonian Dynamics?

It has long been observed that bright matter in spiral galaxies moves more quickly than Newton's law predicts. Such 'flat' rotation curves are what we would expect if the galaxies were spherical rather than disc shaped. Since we see no matter surrounding the bright discs of spiral galaxies, the matter would have to be non-luminous, 'dark' matter. Or does it?

The need for non-detectable, invisible mass arises if we want to hold onto our cherished notions of Newtonian dynamics. In the past there have been several attempts to modify Newtonian dynamics [Fin63, Toh83, Mil83] so as to avoid the need of introducing galactic-scale dark matter. The introduction of a $1/r$-force law was used in an attempt to explain the flat rotational curves at large distances which are a characteristic feature of spiral galaxies [Mil83], and the same law was found to be able to stabilize a cold stellar disc in a n-body galaxy model [Toh83].

An exponential decrease in the surface density was assumed to obey an exponential law, $\sigma(r) = \sigma(0)e^{-r/r_d}$, like in the Debye-Hückel theory, where the optical scale length $r_d = 2.7$ kpc, and $\sigma(0)$ is the central surface density [KK87]. It was also realized that a modified force law that includes a repulsive Yukawa term [Lav95] could describe a range of galaxy rotation curves [San86]. We will argue, in this section, that *these modifications are no more drastic than the Debye-Hückel modification of Coulomb's law in the explanation of ion shielding through a nearest neighbor distribution* [Lav95].

An $1/r$-force law would also be in conflict with Birkhoff's theorem that a spherically symmetrical body cannot radiate in general relativity, as well as any metric accommodating such a force law would not be a solution to Einstein's equations. However, Einstein's equations require all test particles to follow geodesics without any interaction or randomness. Apart from the fact that such a mechanical idealization would be incompatible with a thermodynamic description involving both a finite pressure and temperature, it would also lead to incongruities in attempting to explain galactic interactions with a cosmological model that assumes a universe which is both isotropic and homogeneous.

An inverse-square law has been well-tested over length scales the size of the solar system, but there is no justification for supposing that it holds over 14 magnitudes larger than this length scale! Even the constant rotational speed of galaxies found at distances of the order of $r_0 \approx 10-20$ kpc makes it necessary to modify the viral theorem,

$$v = \left(\frac{GM}{r}\right)^{1/2},$$
(1.53)

at scales $r \gg r_0$ for it fails to satisfy the infrared, Tully-Fisher law [TF77].

The Tully-Fisher is an empirical law that says at such distances the rotational velocity should not vary with the total luminosity of the galaxy,

L, as $v \propto L^{1/2}$, as (1.53) would imply, but rather as

$$v \propto L^{1/4}. \tag{1.54}$$

Since $L \propto M$, (1.54) rules out the classical viral theorem, (1.53). The need to modify (1.53) was apparent, and several suggestions were forthcoming.

What needs to be modified — the inverse-square law or acceleration? Some authors [AEWT90, Lib92] contend that beyond a certain distance the inverse-square law should transform into an inverse-linear law, while others [MdB98] insist that the acceleration must be modified. The reason for the latter is that the surface density is also proportional to the acceleration, $\Sigma_0 = a/G$, and the surface density reflects the surface brightness.

Small accelerations should be related to low-surface-brightness (LSB) objects which should exhibit a larger mass discrepancy than high-surface-brightness (HSB) galaxies. In fact, the centripetal acceleration of LSB galaxies can reach astonishing low values of 10^{-13} cm/s^2.

MOND, or modified Newtonian dynamics, proposed by Milgrom [Mil83] addressed the latter. Using a single parameter, a_0, which was empirically determined to be $a_0 = 1.2 \pm 0.2 \times 10^{-8}$ cm/s^2, or about 100 billion times smaller than the acceleration due to gravity at the earth's surface, could be used to account for the rotation curves of spiral galaxies without the need of assuming dark matter.

The critical value of the acceleration has been proposed as a universal constant of nature which separates regimes for which

$$a = \frac{GM}{r^2},$$

for $a \gg a_0$, from those where

$$a = \frac{(GMa_0)^{1/2}}{r}, \tag{1.55}$$

for $a \ll a_0$. This critical value of the acceleration has been proposed as another universal constant of nature, being the same order of magnitude as cH, the product of the speed of light and Hubble's constant. That is, if a particle will accelerate at a_0, it will reach the speed of light in Hubble's time, $1/H$.

The force law, (1.55), has become a $1/r$ law below a critical acceleration, a_0. It accounts for most of the observed galaxy rotation curves, and reproduces the observed luminosity-velocity relationship in spiral galaxies that are described by the Tully-Fisher law — a distance indicator law — which was originally expressed in terms of the neutral hydrogen line widths and absolute luminosity.

When Milgrom proposed MOND back in 1983, there was already sufficient data on HSB galaxies that reduced it to a parameter fitting exercise [Wor]. Fifteen years were to pass before large populations of LSB galaxies were discovered by observing the 21-cm line. Large discrepancies were found to be present that MOND predicted while dark matter remained completely silent. And, in fact, MOND faired much better than elusive dark matter [KGBS91], even though dark matter had three adjustable parameters while MOND had only one, the ratio of baryonic matter to photons.

Another force law was subsequently proposed by Sanders [San90],

$$a = \frac{GM}{r^2} + \frac{(GMa_0)^{1/2}}{r}, \tag{1.56}$$

which, at large scales, gave the constant rotational velocity,

$$v = (GMa_0)^{1/4}, \tag{1.57}$$

agreeing with the Tully-Fisher law, (1.54).

However, there is absolutely no need to modify Newtonian dynamics, if Debye-Hückel screening is taken into account. Due to a screening of masses about a central mass out to distances of the order of r_0, the Debye-Hückel potential is

$$\Phi(r) = A \frac{e^{-r/r_0}}{r}, \tag{1.58}$$

where A is a constant. The acceleration is the negative gradient of (1.58) which is

$$a = \frac{A \cdot e^{-r/r_0}}{r} \left(\frac{1}{r} + \frac{1}{r_0} \right). \tag{1.59}$$

Setting $A \cdot e^{-r/r_0} = GM$ and $GM/r_0^2 = a_0$ gives back (1.56). With the above cited value for a_0, and $r_0 \approx 10$ kpc, or 30.8×10^{21} cm, the mass M would be about 1.7×10^{44} gm, which is of the order of what Zwicky, back in 1937, estimated as an average mass per galaxy from the entire Coma cluster. He counted roughly 1000 galaxies in the cluster, and found that the average mass per galaxy would be greater than 9×10^{43} g.

The exponential decrease in the central mass with distance is a consequence of the shielding by nearby masses. In fact, (1.59) was proposed earlier by Kuhn and Kruglyak [KK87], but for some unknown reason it was considered to be a failure at lengths $r \gg r_0$ [Lib92].

Closer to the true statistical nature of the Tully-Fisher law was the proposal made by Liboff [Lib92] in which he wrote the acceleration as the geometric mean,

$$a = a_0^{1-p} \left(\frac{GM}{r^2} \right)^{p}, \qquad (1.60)$$

where the parameter p is given by

$$p = \frac{1}{2} \left(1 + e^{-(r/r_0)^2} \right). \qquad (1.61)$$

This form of p was chosen so that it would give Newton's law in the region $r \ll r_0$, and Milgrom's expression, (1.55), in the region $r \gg r_0$. However, (1.61) is not a probability distribution, as can easily be seen by taking its derivative, which must be a positive quantity if it is to coincide with a probability density.

Rather, (1.60) should be replaced by

$$a = \left(\frac{GM}{r^2} \right)^{1-p} a_0^p, \qquad (1.62)$$

and (1.61) should be replaced by the Rayleigh, or nearest neighbor, distribution [Lav95]:

$$p(r) = 1 - e^{-4\pi n r^3/3}, \qquad (1.63)$$

where $4\pi n/3 = 1/r_0^3$ since n is the specific volume. As $r \to 0$, (1.62) goes into the Newtonian law, while as $r \to \infty$ it gives a constant acceleration a_0. At $r = \sqrt[3]{\ln 2}\, r_0$, signaling equal probabilities for Newton's law and constant acceleration, (1.62) gives Milgrom's expression, (1.55). In other words, it is equally likely to attribute an observation of acceleration to be due to a Newtonian cause as it is to one of constant acceleration.

This is basically different from the law (1.59), which asserts that the Milgrom expression, (1.55), is the asymptotic form of the acceleration for $r \to \infty$. In other words, the observation of the flattening of the rotation curves, implying $v = $ const. at large r, which spurned the idea of MOND, should be an intermediary in the asymptotic approach to one of constant acceleration at very large distances.

Milgrom's law is rather peculiar insofar as it does not correspond to any molecular interaction involving ions, dipoles or quadrupoles. The nearest neighbor distribution, (1.63) can be converted into a Fréchet distribution for the largest value of the force [Lav95],

$$p(F) = e^{(F_0/F)^{2\nu-3}}, \tag{1.64}$$

through the inverse power law

$$F(r) = \frac{C_\alpha}{r^\alpha}, \tag{1.65}$$

where F_0 is the 'normal' field [Lav95], and C_α is a constant. In terms of the exponent, ν, $2\nu - 3$ represents half the statistical degrees of freedom of the system, and in order that the three dimensional nearest neighbor distribution result, (1.63), it is necessary to impose the condition, $\alpha(2\nu - 3) = 3$.

For Newton's law, $\alpha = 2$, and $\nu = 9/4$ so that $2\nu - 3 = 3/2$ indicates that there are 3 degrees of freedom. In contrast, for a dipole interaction, $\alpha = 3$ and $\nu = 2$ so that there are 2 statistical degrees of freedom. Rather, for the Milgrom force law, (1.55), $\alpha = 1$, and the number of statistical degrees of freedom are 6 — double those of a system with 3 degrees of freedom.

The acceleration, (1.62), represents the geometric average of the normal, inverse square law and a constant acceleration. This is the least acceleration possible. It can be thought of as a limiting value of the s mean,

$$\mathcal{M}_s = \left\{ (1-p) \left(\frac{GM}{r^2} \right)^s + pa_0^s \right\}^{1/s} \tag{1.66}$$

in the limit as $s \to 0$. And since the means are monotonic increasing functions of their order, $\mathcal{M}_s > \mathcal{M}_r$ for $s > r$, the geometric mean, (1.62), is the smallest mean possible for a positive value of s.

MOND is no more radical than the Debye-Hückel theory of electrolytes. Why it is in difficulty with general relativity is that it does not treat the hydrodynamics of perfect fluids, which is a sorry excuse for modeling the evolution of galaxies and their interactions. The assertion that MOND cannot be tested against general relativistic effects like gravitational lensing is a lack of appreciation of what it achieves. The magnitude of gravitational lensing is proportional to the deviation of the angular momentum of a particle from a constant of the motion in the field of a more massive body [Lav11]. It has nothing to do with a modification of Newton's law of gravity.

Criticism has been lodged against MOND because "it has never been developed into a complete theory" [Liv00], and "it must remain as a last resort" [Rub96]. According to Rubin, "Most astronomers prefer to accept a universe filled with dark matter rather than to alter Newtonian gravitational theory." If most astronomers want to remain in the dark ages it is their prerogative.

Newtonian gravitation describes the attraction between two bodies, but it fails to account for the gravitational shielding by many bodies which is described *statistically*, in this case by a Holtzmark distribution [Lav95]. It is in this realm that deterministic laws must give way to statistical ones, and seen in this light, there is nothing repugnant about MOND, except contradicting 'consensus' cosmology of a flat, accelerating universe that has undergone a period of inflation since its birth in a hot big bang. To people like Livio [Liv00], an accelerating universe is so beautiful that it 'has' to be true. This is reminiscent of Boltzmann's remark that he would leave aesthetics to tailors and cobblers, which has no place in the description of physical phenomena.

Bibliography

[AEWT90] M. J. Disney, A. E. Wright and R. C. Thomson. Universal gravity: was Newton right? *Proc. ASA*, 8(4):334–338, 1990.

[AGP05]   M. Joyce, A. Gabrielli, F. S. Labini and L. Pietronero. *Statistical Physics for Cosmic Structures*. Springer-Verlag, Berlin, 2005.

[BA98]   C. F. Bohren and B. A. Albrecht. *Atmospheric Thermodynamics*. Oxford University Press, New York, 1998.

[Bal07]   A. Balbi. *The Music of the Big Bang*. Springer, New York, 2007.

[BH76]   H. P. Baltes and E. R. Hilf. *Spectra of Finite Systems*. Bibliographisches Institut AG, Zürich, 1976.

[BH83]   C. F. Bohren and D. R. Huffman. *Absorption and Scattering of Light by Small Particles*. Wiley-Interscience, New York, 1983.

[CBZ91]   L. Danese, C. Burigana and G. De Zotti. *Astron. Astrophys.*, 246:49, 1991.

[Cha50]   S. Chandrasekhar. *Radiative Transfer*. Oxford University Press, Oxford, 1950.

[Col77]   S. Coleman. The fate of the false vacuum: Semiclassical theory. *Phys. Rev. D*, 15:2929–2936, 1977.

[Cot99]   A. Cotton. The present state of Kirchhoff's law. *Astrophys. J*, 9:237–268, 1899.

[Dic46]   R. H. Dicke. The measurement of thermal radiation at microwave frequencies. *Rev. Sci. Instr.*, 17:268–275, 1946.

[DL01]   T. M. Davis and C. H. Lineweaver. *Superluminal recession velocities*. Technical Report. arXiv:astro-ph/0011070v2, January 2001.

[FHN00]   G. Burbidge, F. Hoyle and J. V. Narlikar. *A Different Approach to Cosmology.* Cambridge University Press, Cambridge, 2000.

[Fin63]   A. Finzi. On the validity of Newton's law at long distance. *Mon. Not. R. Astron. Soc.*, 27:21–30, 1963.

[GS81]   H. S. Goldberg and M. D. Scadron. *Physics of Stellar Evolution and Cosmology.* Gordon and Breach, New York, 1981.

[Gut81]   A. Guth. Inflationary universe: A possible solution to the horizon and flatness problems. *Phys. Rev. D*, 23:347–356, 1981.

[GY89]   R. M. Goody and Y. L. Yung. *Atmospheric Radiation.* Oxford University Press, New York, 2nd edition, 1989.

[Har00]   E. Harrison. *Cosmology.* Cambridge University Press, Cambridge, 2nd edition, 2000.

[HW97]   W. Hu and M. White. *A CMB polarization primer.* Technical Report. arXiv:astro-ph/9706147, 1997.

[IS75]   A. F. Illarionov and R. A. Sunyaev. *Soviet Astron. J.*, 18:691, 1975.

[KGBS91]   A. H. Broeils, K. G. Begeman and R. H. Sanders. Extended rotation curves of spiral galaxies: dark halos and modified dynamics. *Mon. Not. R. Astron. Soc.*, 249:523–537, 1991.

[KK87]   J. Kuhn and L. Kruglyak. Non-Newtonian force and the invisible mass problem. *Astrophys. J.*, 313:1–12, 1987.

[KT90]   E. W. Kolb and M. S. Turner. *The Early Universe.* Addison-Wesley, Reading MA, 1990.

[Lav85]   B. H. Lavenda. *Nonequilibrium Statistical Thermodynamics.* Wiley, Chichester, 1985.

[Lav91]   B. H. Lavenda. *Statistical Physics: A probabilistic approach.* Wiley-Interscience, New York, 1991.

[Lav95]   B. H. Lavenda. *Thermodynamics of Extremes.* Horwood, Chichester, 1995.

[Lav09]   B. H. Lavenda. *A New Perspective on Thermodynamics.* Springer, 2009.

[Lav11]   B. H. Lavenda. *A New Perspective on Relativity: An Odyssey in Non-Euclidean Geometries.* World Scientific, Singapore, 2011.

[Lib92]   R. L. Liboff. Generalized Newtonian force law and hidden mass. *Astrophys. J.*, 397:L71–L73, 1992.

[Liv00]   M. Livio. *The Accelerating Universe.* John Wiley, New York, 2000.

[LL58]   L. D. Landau and E. M. Lifshitz. *Statistical Physics*. Pergamon
          Press, Oxford, 1958.

[LL59]   L. D. Landau and E. M. Lifshitz. *Fluid Mechanics*. Pergamon
          Press, Oxford, 1959.

[McK40]   A. McKellar. Evidence for the molecular origin of some hith-
          erto unidentified interstellar lines. *Pub. Astr. Soc. Pacific*, 52:
          187–192, 1940.

[MdB98]   S. S. McGaugh and W. J. G. de Bloke. Testing the hypothesis
          of modified dynamics with low surface brightness galaxies and
          other evidence. *Astrophys. J.*, 499:66–81, 1998.

[Mil83]   M. Milgrom. A modification of the Newtonian dynamics as a
          possible alternative to the hidden mass hypothesis. *Astrophys.
          J.*, 270:365–370, 1983.

[Muk05]   V. Mukhanov. *Physical Foundations of Cosmology*. Cambridge
          University Press, Cambridge, 2005.

[Nan79]   G. P. Nanos. Polarization of blackbody radiation at 3.2 cm.
          Astrophys. J., 232:341–347, 1979.

[Par95]   R. B. Partridge. *3K: The Cosmic Microwave Background Radi-
          ation*. Cambridge University Press, Cambridge, 1995.

[Pea99]   J. A. Peacock. *Cosmological Physics*. Cambridge University
          Press, Cambridge, 1999.

[Pee80]   P. J. E. Peebles. *Large Scale Structure of the Universe*. Prince-
          ton University Press, Princeton NJ, 1980.

[Pee93]   P. E. J. Peebles. *Principles of Physical Cosmology*. Princeton
          University Press, Princeton NJ, 1993.

[Pla14]   M. Planck. *Theory of Heat Radiation*. Blackiston & Son,
          Philadelphia, PA, 1914.

[Ree68]   M. Rees. Polarization and spectrum of the primeval radiation
          in an isotropic Universe. *Astrophys. J.*, 153:L1–L5, 1968.

[Rub96]   V. C. Rubin. *Bright Galaxies Dark Matters*. Springer-Verlag,
          New York, 1996.

[San86]   R. H. Sanders. Alternatives to dark matter. *Mon. Not. R.
          Astron. Soc.*, 223:539–555, 1986.

[San90]   R. H. Sanders. Mass discrepancies in galaxies: dark matter and
          alternatives. *Astron. & Astrophys. Rev.*, 2:1–28, 1990.

[SZ98]   U. Seljak and M. Zalderriaga. *Polarization of the microwave
          background: Statistical and physical properties*. Technical
          Report. arXiv:astro-ph/980501, May 1998.

[TF77]   R. B. Tully and J. R. Fisher. A new way of determining distances to galaxies. *Astron. & Astrophys.*, 54:661–673, 1977.

[Toh83] J. E. Tohline. *IAU Symposium: Internal Kinematics and Dynamics of Galaxies*, page 205. Reidel, Dordrecht, NL, 1983.

[Wei60] M. A. Weinstein. On the validity of Kirchhoff's law for a freely radiating body. *Am. J. Phys.*, 28:123–125, 1960.

[WH97]   M. White and W. Hu. The Sachs-Wolfe effect. *Astron. Astrophys.*, 321:8–9, 1997.

[Wor]   B. Worraker. *MOND over dark matter*. Technical report, creation. com/mond-over-dark-matter.

Chapter 2

Is the Universe Hydrodynamic?

2.1 Fluid Dynamics of General Relativity

In 1934 McCrea and Milne [MM34] made the startling discovery that Newtonian gravitation, when applied to a uniform universe, yields the same results as general relativity. Or does it?

Armed with the continuity equation alone, Milne and McCrea single handedly derived the Friedmann-Lemaître equations for the expansion rate of the universe at zero pressure. That they couldn't take pressure into account, nor could Harrison [Har00] nor Pebbles [Pee93], contrary to what has been claimed, was already a tell-tale sign that something was very wrong.

And that something was *the confusion of a fluid velocity with the time derivative of distance*. In the very different branch of irreversible thermodynamics, Onsager knew he could only derive his famous reciprocal relations for coupled transport coefficients from the principle of microscopic reversibility if the fluxes were time derivatives of extensive variables [Lav78]. This necessarily excluded all cross-symmetries between coupled flows and forces from transport phenomena.

In other words, McCrea and Milne confused a material velocity v for the flow of matter with the time rate of change of a scale factor, S. Moreover, it is manifestly apparent that the Friedmann-Lemaître equation for the acceleration of the scale factor could never be compatible with Euler's equation because the pressure enters linearly into the former while it is the gradient

77

that is responsible for fluid acceleration in the latter. This has led to some of the most ridiculous claims of what pressure can do. "Pressure (stress) is a form of energy density and is therefore a source of gravity" [Har00].

Although there may be a relation between pressure and density, pressure is not a source of gravity, if Euler has something to say about it. And, yet another ludicrous claim is: "Pressure of radiation exerts its own gravitational field, thereby increasing the amount of gravity acting." The pressure of radiation is so minute that who would ever think of it as creating its own gravitational field?

The starting point for the Milne-McCrea analysis is Newton's equation,

$$\frac{dv}{dt} = -\frac{M(r)}{r^2}, \tag{2.1}$$

in units where $G = c = 1$, which they observed may be written as

$$\frac{\partial v}{\partial t} + v\frac{\partial v}{\partial r} = -\frac{4\pi}{3}\rho r, \tag{2.2}$$

where the density ρ can only be a function of time.

We can introduce the pressure p by writing Newton's equation (2.1) as Euler's equation,

$$\frac{dv}{dt} = -\frac{1}{\rho}\frac{dp}{dr}. \tag{2.3}$$

Then, equating the right-hand sides of (2.1) and (2.3) gives

$$dp = \frac{M}{r^2}\rho dr, \tag{2.4}$$

which is *not* the equation of hydrostatic equilibrium. So something is definitely amiss, and that something has equated a particle acceleration in (2.1) with a hydrodynamic acceleration in (2.3). The two are not one and the same thing, as the wrong sign in (2.4) bears witness to that.

Euler's equation in the presence of a gravitational field is

$$\frac{dv}{dt} = -\frac{1}{\rho}\frac{dp}{dr} - \frac{M}{r^2}. \tag{2.5}$$

If the fluid is at rest, $v = 0$, (2.5) reduces to the equation of hydrostatic equilibrium,

$$dp = -\frac{M}{r^2}\rho dr, \tag{2.6}$$

which differs from (2.4) by a (mere) sign.

Now let us see what happens when we try to use the Friedmann-Lemaître equation in the presence of a pressure,

$$\ddot{r} = -\frac{4\pi}{3}\left(\rho + 3p\right)r, \tag{2.7}$$

to eliminate the last term in Euler's equation written in terms of the density,

$$\frac{dv}{dt} = -\frac{1}{\rho}\frac{dp}{dr} - \frac{4\pi}{3}\rho r. \tag{2.8}$$

We then get

$$\frac{dv}{dt} = -\frac{1}{\rho}\frac{dp}{dr} + \ddot{r} + 4\pi pr. \tag{2.9}$$

So if $v = \dot{r}$, it would imply that

$$4\pi\rho = \frac{1}{r}\frac{d}{dr}\ln p,$$

which is not Poisson's equation, nor can it be transformed into it on the condition of hydrostatic equilibrium, (2.6).

Rather, the correct equation is

$$4\pi\rho = -\frac{1}{r^2}\frac{d}{dr}\left(\frac{r^2}{\rho}\frac{dp}{dr}\right),$$

which can be transformed into Poisson's equation,

$$4\pi\rho = \frac{1}{r^2}\frac{d}{dr}\left(r^2\frac{d\Phi}{dr}\right), \tag{2.10}$$

by (2.6), where Φ is the Newtonian gravitational potential. Consequently, $v \neq \dot{r}$ even when the pressure is the same at every point in the fluid because it enters Einstein's equation (2.7) as p itself, and not as its gradient as in Euler's equation (2.8). Moreover, the discrepancy in sign means that pressures acts as its negative in the Friedmann-Lemaître equation (2.7) which has falsely been held responsible for such things as blowing up universes.

So it is not as Harrison claims that the Newtonian picture of a sphere expanding in flat Euclidean space, and Einstein's picture of it expanding in a curved space, "despite their basic differences, yield identical equations." It is not that the "Newtonian picture breaks down when the pressure is not small compared with the energy density ρ," for ρ is the rest energy density and it is not comparable to the pressure. The pressure is, however, comparable to the internal energy density which does not appear in Einstein's equations.

Gravity causes both a solid and fluid particles to accelerate, but Euler's equation claims that is the *gradient* of the pressure that causes the rate of change of a given fluid particle as it moves about in space. Gravity is a force, and so too is the gradient of the pressure. In other words, pressure is stress, but it is the change in the stress that causes motion. And this is unlike gravity which causes motion even when it is uniform.

In the weak (Newtonian) gravitational field, the time component of the metric tensor is $g_{00} = -1 - 2\Phi$. The Newtonian gravitational potential Φ, in this limit, satisfies Poisson's equation, (2.10). Rather, if Einstein's equation, (2.7), was true we would have to generalize (2.10) to

$$\frac{1}{r^2}\frac{d}{dr}\left(r^2\frac{d\Phi}{dr}\right) = 4\pi(\rho + 3p), \qquad (2.11)$$

making pressure a source of gravity even in a weak gravitational field! Pressure is not a source of gravity no matter what limit is considered so (2.11) is wrong.

We, therefore, do not share the opinion of Callan, Dicke and Peebles [CCP65] that

> Long ago, McCrea and Milne showed that Newtonian mechanics was capable of describing the expansion of the universe. However, it does not seem to have been generally recognized that such a classical treatment was not the dynamics of a crude classical model of the universe but was rather a completely correct treatment of the real universe.

2.2 Raychaudhuri's Identity: Are Singularities Real?

All models of the universe that are based on Einsteinian relativity are hydrodynamic in nature. They are invariably based on the Robertson-Walker metric which predicts both a homogeneous and isotropic evolution. Since the early days of general relativity, the practitioner has been dogged by the appearance of singularities in an otherwise uniform flow field. Were these singularities to be taken seriously? Einstein believed not, but the present day consensus has overruled him.

Treating geodesics as lines of fluid motion, interest was focused on what type of criteria would determine whether families of geodesics, called congruences, would bunch up and form a singularity, analogous to the convergence of rays in optics to form a caustic. These later became known as focusing theorems; a singularity would always imply the focusing of geodesics, but the converse is not necessarily true [LL51].

The usual description of fluid motion is through a velocity field, and the characterization of the motion in terms of a velocity potential, and a stream function would imply a grid map on which to view the motion. The velocity is normal to surfaces of constant velocity potential, and always parallel to the stream surface.

In order to make contact with the Ricci tensor, and from there the Einstein equations, Raychaudhuri [Ray79] found it necessary to go one derivative higher and consider the gradient of the velocity field. The gradient of the velocity field, being a second order tensor, can be split up into a trace, a symmetric and an antisymmetric traceless components. These define the principal stresses: the expansion, shear, and torsion.

The Raychaudhuri identity is derived from the definition of the Riemann tensor,

$$\nabla_\alpha \nabla_\beta u^\mu - \nabla_\beta \nabla_\alpha u^\mu = \frac{1}{2} R^\mu_{\nu\beta\alpha} u^\nu, \tag{2.12}$$

where u^μ is an arbitrary vector, which we will subsequently identify as the four velocity, and the indices run over time and the three spatial dimensions. If (2.12) is contracted by setting $\mu = \beta$, and multiplied through by u^α there results

$$\frac{d\Theta}{dt} + \frac{1}{3}\Theta^2 + 2\sigma_{\alpha\beta}\sigma^{\alpha\beta} - 2\omega_{\alpha\beta}\omega^{\alpha\beta} - \nabla_\beta u_\alpha u^\beta = R_{\alpha\beta} u^\alpha u^\beta, \tag{2.13}$$

where $\Theta = \nabla_\alpha u^\alpha$ is the expansion, and $d\Theta/dt = u^\alpha \partial\Theta/\partial x^\alpha$ is its total time derivative. The shear is

$$\sigma_{\alpha\beta} = \nabla_\beta u^\alpha + \nabla_\alpha u^\beta - \frac{1}{3} h_{\alpha\beta}\Theta - u^\lambda \nabla_\lambda u_\alpha u^\beta, \tag{2.14}$$

with $h_{\alpha\beta} = g_{\alpha\beta} + u_\alpha u_\beta$ as the transverse part of the metric $g_{\alpha\beta}$, and

$$\omega_{\alpha\beta} = \nabla_\beta u^\alpha - \nabla_\alpha u^\beta - u^\lambda \left(u^\beta \nabla_\lambda u_\alpha - u^\alpha \nabla_\lambda u_\beta \right), \tag{2.15}$$

is the torsion. $R_{\alpha\beta}$ is the Ricci tensor. Both shear, (2.14), and torsion, (2.15), are normal to the flow, u^β, since

$$\sigma_{\alpha\beta} u^\beta = \omega_{\alpha\beta} u^\beta = 0. \tag{2.16}$$

2.3 Evolution of Congruences: Can Negative Pressure Cause Expansion?

The tensorial identity (2.13) is converted into an equation when the Ricci tensor is replaced by

$$R_{\alpha\beta} = 8\pi \left(T_{\alpha\beta} - \frac{1}{2}g_{\alpha\beta}T \right), \qquad (2.17)$$

where $T_{\alpha\beta}$ is the matter-energy tensor of the Einstein equations in units where $G = c = 1$. For irrotational motion, (2.15) vanishes.

Considering the evolution of a family of geodesics, referred to as congruences, the criterion for them to converge is when the 'strong energy' condition,

$$R_{\alpha\beta}u^\alpha u^\beta \geq 0, \qquad (2.18)$$

holds. The Raychaudhuri equation then implies

$$\dot{\Theta} = -\frac{1}{3}\Theta^2 - \sigma_{\alpha\beta}\sigma^{\alpha\beta} - R_{\alpha\beta}u^\alpha u^\beta \leq 0, \qquad (2.19)$$

so that the expansion must decrease. The strong energy condition, (2.18), is commonly attributed to gravity, and its attractive force is what makes the family of geodesics converge.

The two most common metrics to which the Raychaudhuri equation has been applied is the Schwarzschild metric,

$$d\tau^2 = (1 - 2M/r)dt^2 - \frac{dr^2}{1 - 2M/r} - r^2 d\Omega^2, \qquad (2.20)$$

where $d\Omega^2$ is the line element for a two sphere, for which the Ricci tensor vanishes even though there is mass M present, and the Robertson-Walker metric,

$$ds^2 = -d\tau^2 + S^2(\tau)d\gamma^2, \qquad (2.21)$$

where τ is the proper time along the world lines, and $d\gamma^2$ is a line element of a space of zero, unit positive or unit negative curvature. If the Ricci tensor does not vanish, and if the evolution is that of a perfect fluid, the matter-energy tensor is

$$T_{\alpha\beta} = \varrho u_\alpha u_\beta + p\, h_{\alpha\beta}, \qquad (2.22)$$

where ϱ is a 'density,' and $h_{\alpha\beta} = g_{\alpha\beta} + u_\alpha u_\beta$ is the projection on the plane orthogonal to u_α, where the four velocities satisfy

$$g^{\alpha\beta}u_\alpha u_\beta = -1. \qquad (2.23)$$

The hydrostatic pressure, p, has now been upgraded to a 'cosmological' pressure which supposedly accounts for the interactions of galaxies as if they were mere molecules.

A perfect fluid is inviscid and has constant density; it is characterized by the absence of shear stresses between the fluid layers. The components of the velocity vector is derivable as the gradient of a velocity potential,

$$u_\tau = \frac{d\tau}{dt},$$

where the surfaces $\tau = $ const. are both homogeneous and isotropic, representing three surfaces of constant curvature.

Defining a time t according to [Haw66]

$$\frac{dt}{d\tau} = \frac{1}{S},$$

the Robertson-Walker metric, (2.21), can be put into the conformal form

$$ds^2 = S^2(t)(-dt^2 + d\gamma^2).$$

The square root of the determinant of the metric is $\sqrt{-g} = S^2$, and the velocity is $u_\tau = S$.

The expansion, therefore, is

$$\Theta = \frac{1}{\sqrt{-g}}\left(\sqrt{-g}u_\tau\right)_{,\tau} = 3\frac{dS}{d\tau} = 3\frac{dS}{dt}\frac{dt}{d\tau} \equiv 3\frac{\dot{S}}{S}, \qquad (2.24)$$

where the comma stands for differentiation with respect to the variable that follows it. The physical significance of (2.24) can be uncovered by multiplying the numerator and denominator by S^2; for then we obtain

$$\Theta = \frac{\dot{V}}{V}, \qquad (2.25)$$

which expresses the expansion in terms of the fractional rate of change of the volume, V. In fact, (2.25) can be taken as the definition of the expansion.

The singularity $S = 0$ is the most prominent feature of the Robertson-Walker metric. It occurs in all models in which the strong energy condition, $\varrho + 3p \geq 0$, holds, since shear, (2.14), vanishes for a perfect fluid, and the Raychaudhuri equation, (2.19), reduces to

$$\dot{\Theta} = -\frac{1}{3}\Theta^2 - 4\pi(\varrho + 3p) \leq 0. \qquad (2.26)$$

Why would anyone think of applying the convergence of congruence criteria to the Schwarzschild metric (2.20) which has nothing in common with hydrodynamic motion? For the Schwarzschild metric, the radial component of the velocity is $u_r = \sqrt{2M/r}$, and, consequently, the expansion is [Poi04]

$$\Theta = \pm\frac{1}{\sqrt{-g}} \left(\sqrt{-g}u_r\right)_{,r} = \pm\frac{3}{2}\sqrt{\frac{2M}{r^3}}, \qquad (2.27)$$

where the determinant $\sqrt{-g} = r^2 \sin\theta$, and the plus and minus has been added for out going and incoming geodesics.

It will begin to look more like fluid behavior when we introduce the density, $M = 4\pi\varrho r^3/3$, for then $u_r = \sqrt{8\pi\varrho/3}\, r$. Anticipating that we must obtain the first of Einstein's field equation, we observe that (2.27) must be doubled. With this correction (2.27) becomes

$$\Theta^2 = 24\pi\varrho. \qquad (2.28)$$

Equating this with (2.24) gives the first of the field equations

$$8\pi\varrho = 3\frac{\dot{S}^2}{S^2}, \qquad (2.29)$$

for a flat universe and zero cosmological constant.

Even if we wanted to, we have no rule, nor motivation, for introducing additional terms in (2.29). Einstein referred to the introduction of his cosmological constant as "the biggest blunder of his life." More recently, Hawking has referred to the prediction of an event horizon for a black hole as his "life's biggest blunder." These are not blunders in the sense of mistakes, but, rather, are in vane attempts to patch up a flawed theory.

Introducing (2.24) and (2.27) into the Raychaudhuri equation (2.26) then gives the second field equation,

$$8\pi p = -\frac{2S\ddot{S} + \dot{S}^2}{S^2} = -\frac{4}{3}\left(\frac{1}{3}\Theta^2 + \frac{1}{2}\dot{\Theta}\right). \qquad (2.30)$$

Pressure is *negative* hydrostatic stress, and because the latter is compressive, the pressure is positive. Negative pressure can exist, but with restricted stability, since it causes a spontaneous contraction of the body [LL].

Expression (2.30) is *not* a relation between the pressure and the negative of the average of the principal stresses. Moreover, the 'focusing' theorem requires $\dot{\Theta} < 0$ for which a family of geodesics will tend to contract which, according to (2.30), will cause positive pressure causing an expansion of a body, and not a negative pressure which would cause a body to spontaneously contract. There are no hydrodynamic analogs to either (2.28) or (2.30), and yet, it is, indeed, strange why they are almost universally accepted.

Again, the cosmological constant has be put in by hand, and it will affect the magnitude of the pressure. Early on in the history of general relativity, de Sitter found that the cosmological constant can blow up an otherwise empty universe. It is at this point where, when questioned by a Dutch reporter to a local newspaper as to what makes the universe swell up, de Sitter responded "that is done by Lambda, another answer cannot be given."

Moreover, just by introducing (2.28) into (2.26), gives the 'weak' energy condition,

$$\dot{\Theta} = -12\pi(\varrho + p) \leq 0. \tag{2.31}$$

Rather, had we differentiated the corrected equation (2.27) at constant mass we would have obtained

$$\dot{\Theta} = -\frac{9}{2}\sqrt{\frac{2M}{r^5}}\, u_r = -9\frac{M}{r^3} = -12\pi\varrho. \tag{2.32}$$

Comparison with (2.31) shows that the pressure has vanished and we are dealing with dust! And introducing (2.28) into (2.32) shows that the expansion is decreasing with time since $\Theta = 2/t$.

Landau's name is sometimes attached to the Raychaudhuri equation [KS07] because Landau also addressed the problem of focusing, albeit in a rather camouflaged way [LL51]. Landau was interested in calculating the components of the Ricci tensor by separating off the space part of the metric. He claimed that a reference system could always be found in which $g_{00} = -1$, and $g_{0\alpha} = 0$.

In the present case, Landau would have found for the time component of the Ricci tensor

$$R_{00} = \dot{\Theta} + \frac{1}{3}\Theta^2 = -12\pi\varrho + 8\pi\varrho = -4\pi\varrho.$$

That this quantity is negative implies, according to (2.18), the strong energy condition, but Landau does not refer to it as such. Through Landau's approach it becomes evident that the space component of the metric determines the dynamics of the flow, and there is no need for the metric to reduce to the flat metric of special relativity in the absence of gravitation. Moreover, since the first equality holds in general, Landau also shows that the additional components of shear and torsion do not enter into R_{00}.

The vanishing of the pressure, (2.30), implies that $\dot{S}^2 S = \text{const.}$, which becomes $\Lambda S^2 = 2S\ddot{S} + \dot{S}^2 + k$ when we throw in the cosmological constant Λ, and the normalized curvature $k = 0, \pm 1$. Expression (2.30) shows that the pressure is negative for an accelerating universe or, for that matter, even an inertial one. Moreover, it is time dependent; this, indeed, is a far cry from the Planck's invariant pressure of the special theory.

2.4 Mixing Schwarzschild and Robertson-Walker Metrics

Parenthetically, we may illustrate just how confusing things can become by taking an example from Landau and Lifshitz [LL59]. They begin with the Robertson-Walker matter-energy tensor, (2.22), take its covariant derivative, $\nabla_\beta T_\alpha^\beta$, and then contract it with u^α to obtain

$$hu^\beta \nabla_\beta u_\alpha + \frac{\partial p}{\partial x^\alpha} + u_\alpha u^\beta \frac{\partial p}{\partial x^\beta} = 0,$$

claiming that the ordinary derivative must be replaced by the covariant one when gravitational fields are present, where $h = \varepsilon + p$ is the enthalpy density. However, this disregards the Einstein condition for energy and momentum conservation, $\nabla_\beta T_\alpha^\beta = 0$.

In the static case, the only surviving velocity component is $u^0 = 1/\sqrt{-g_{00}}$. The space components are then

$$-hu^0 u_0 \Gamma_{\beta 0}^0 = \frac{1}{2} h \frac{1}{g_{00}} \frac{\partial g_{00}}{\partial x^\beta} = h \frac{\partial \ln \sqrt{-g_{00}}}{\partial x^\beta} = -\frac{\partial p}{\partial x^\beta}.$$

For the Schwarzschild metric, $-g_{00} = 1 - 2M/r$, and in the non relativistic limit, $h \simeq \varrho$, one obtains grad $p = $ grad $2M/r$, the equation of hydrostatic equilibrium.

Landau and Lifshitz have mixed together the Robertson-Walker and the Schwarzschild metrics. The latter has a vanishing mass-energy tensor, and their pressure is of the normal garden variety, having nothing to do with the relativistic pressure in (2.30). They have, moreover, confused the internal energy density in the expression enthalpy density with the rest energy, which, in any event, cannot be considered the non-relativistic limit. The rest energy does not satisfy any thermodynamic relation!

From equations (2.29) and (2.30) it is not difficult to show that the adiabatic condition

$$\frac{d}{dt}\left(\varrho V\right) + p\frac{dV}{dt} = 0, \tag{2.33}$$

is satisfied. In the case of the Schwarzschild metric, (2.32) is a statement that the total mass is time-independent. The condition is that the pressure, (2.30), vanish. The Raychaudhuri equation (2.26) then reduces to

$$\dot{\Theta} = -\frac{1}{3}\Theta^2 - 4\pi\varrho. \tag{2.34}$$

This is none other than $\dot{\varrho} = -\sqrt{24\pi\varrho}\,\varrho$, or the conservation of mass,

$$\dot{\varrho} = -\varrho\Theta, \tag{2.35}$$

when (2.24) is used. The vanishing of the pressure imposes the condition $\acute{S}^2 S = \text{const.}$ There is no generalization of the Raychaudhuri equation (2.34) when ϱ is the mass density, in which case it is a mere identity.

In view of (2.29), ϱ must be interpreted as the rest energy density because the mass density of a perfect fluid is constant. Furthermore, in view of the adiabatic condition (2.33), it must undergo a further interpretation as the internal energy density ε. In place of the conservation of mass (2.35), we now have the condition $\dot{\varepsilon} = -\Theta(\varepsilon + p)$, which upon introducing (2.24) becomes

$$\frac{\dot{\varepsilon}}{\varepsilon + p} = -3\frac{\dot{S}}{S} = \frac{\dot{s}}{s}, \tag{2.36}$$

where s is the entropy density.

The first equality in (2.36) is an identity, and it is equivalent to the adiabatic condition, (2.33). The last equality expresses the adiabatic condition as

$$sS^3 = \text{const.}$$

The first and third equalities are the definition of the inverse absolute temperature.

The equation of continuity, when expressed as the covariant divergence,

$$\frac{1}{\sqrt{-g}}\left(\sqrt{-g}nu_\tau\right)_{,\tau} = 0,$$

is satisfied identically, where

$$n^{-1} = S^3,$$

is the specific volume. So where do the dissipative elements come in?

The fact that the square of the expansion, (2.28), and its time derivative, $d\Theta^2 V/dt = -24\pi p dV/dt$, are functions of the density and pressure is completely foreign to hydrodynamics. These relations follow from the normalization condition, (2.23), which has no counterpart in fluid mechanics.

This has the effect of mixing non dissipative and dissipative elements, which is like mixing apples and bananas.

2.5 Dissipation in Relativistic Fluids

Dissipative terms, in thermodynamics of relativistic fluids, are included by a mass-energy tensor of the form [LL59]:

$$T_\alpha^\beta = h u_\alpha u^\beta + p \delta_\alpha^\beta + \tau_\alpha^\beta, \tag{2.37}$$

where

$$\tau_\alpha^\beta = -\eta \left(\nabla_\beta u^\alpha + \nabla_\alpha u^\beta + u^\lambda \nabla_\lambda u_\alpha u^\beta \right) - \left(\zeta - \frac{2}{3}\eta \right) \Theta h_\alpha^\beta, \tag{2.38}$$

η and ζ are two viscosity coefficients, and $h_\alpha^\beta = \delta_\alpha^\beta + u_\alpha u^\beta$. It is easy to check that (2.38) satisfies (2.16). And because of the condition that the stress must be transverse to the flow, (2.16), the entropy flux density, $s^\alpha \equiv s u^\alpha$ will satisfy

$$\nabla_\alpha s^\alpha = -\frac{\tau_\alpha^\beta}{T} \nabla_\beta u^\alpha,$$

at constant temperature T, and the relation $\varrho + p = Ts$, valid at vanishing chemical potential, has been used.

Even if the shearing stress vanishes, i. e., $\eta = 0$, there is a non-vanishing contribution coming from the bulk viscosity, $viz.$,

$$\nabla_\alpha s^\alpha = \frac{\zeta}{T}\Theta^2 \geq 0. \tag{2.39}$$

Consequently, the Raychaudhuri equation (2.26) is incompatible with the condition of adiabaticity, although it formerly reproduces it.

That is to say that we can write the mass-energy tensor as

$$T_\alpha^\beta = \varepsilon u_\alpha u^\beta + p^\star (\delta_\alpha^\beta + u_\alpha u^\beta),$$

with a new pressure,

$$p^\star = p - \zeta\Theta. \tag{2.40}$$

Both the Einstein equations and the Raychaudhuri equation would formally be the same where the old pressure p would be replaced by the new pressure (2.40).

According to Weinberg [Wei72], the only dissipative effect that can enter the mass-energy tensor for an isotropic, homogeneous expansion is the bulk viscosity. Notwithstanding that it appears in its trace,

$$T_\alpha^\alpha = 3p^\star - \varepsilon.,$$

even if we didn't know what $p^\star$ was, we would see that it satisfies the same 'adiabatic' Einstein equations, and the Raychaudhuri equation (2.26). According to its general relativistic definition (2.30), the pressure is time dependent and is *negative* for an accelerating universe. This is a consequence of the adiabaticity condition (2.33).

In other words, there is nothing in geometry that would enable us to determine the direction of time. Recall that

$$\dot\Theta = 3\frac{S\ddot S - \dot S^2}{S^2},$$

and is, therefore, invariant under time inversion. Anything that we introduce into that equation must also be time invariant. This must also be true of the Einstein equations, (2.17), since the left-hand side is pure geometry. This discourages us from introducing mass-energy tensors like (2.37) into the Einstein equations, (2.17).

The same is true of the Rauchaudhuri equation, (2.26). For if we did introduce a pressure like (2.40), into (2.26), we would find

$$\frac{S\ddot S + \dot S^2}{S^2} + \frac{8\pi}{3}p = 8\pi\zeta\frac{\dot S}{S}. \tag{2.41}$$

The left-hand side of (2.41) is invariant to time inversion, while the right-hand side is not. It is not by chance that both viscosity coefficients are absent in the Rauchaudhuri identity, (2.19), and, moreover, they must still be absent when the Ricci tensor is evaluated with the Einstein equations, (2.17).

The story is different for the shear viscosity, since it cannot be absorbed into a scalar quantity. Now setting the bulk viscosity $\eta = 0$, the stress tensor becomes

$$\tau_\alpha^\beta \nabla_\alpha u^\beta = -\eta\left[\frac{1}{2}\left(\nabla_\alpha u^\beta + \nabla_\beta u^\alpha\right)^2 - \frac{2}{3}\Theta^2 + u^\lambda\nabla_\lambda u_\alpha u^\beta\right]. \tag{2.42}$$

We now appeal to space inversion. The Rauchaudhuri identity (2.19) is invariant under the inversion of the shear stress. The term we would introduce in the mass-energy tensor is (2.37), and inserting it into the Rauchaudhuri equation, (2.19), via the Einstein equations, (2.17), would make the former linear, rather than only quadratic, in terms of the shear stress.

It would appear from (2.42) that dilation would lead to a decrease in the rate of the density of entropy flow. However, this is only one of appearances because the first two terms in (2.42) combine to form a perfect square,

$$\frac{\eta}{2}\left(\nabla_\alpha u^\beta + \nabla_\beta u^\alpha - \frac{2}{3}\delta^\beta_\alpha\Theta\right)^2.$$

From the Raychaudhuri equation, (2.19), we would conclude that the symmetric shear component works in conjunction with dilation to induce a contraction of the congruence; yet, from (2.39) it appears that they have opposite effects on the rate of entropy flow density. Then, there is the bigger question as to what these terms have to do with an adiabatic process governed by (2.33)?

The strong and weak energy conditions, (2.18) and (2.31), respectively, are necessary conditions for the Hawking-Penrose singularity theorem. Hawking and Penrose [HE73] assume that from the observed CMBR there is sufficient energy "on each past directed null geodesic" from our present position in the universe to have them focus so as to develop a caustic at some point in spacetime.

A caustic is a surface of conjugate points separating light and dark regions. The expansion, Θ, must therefore decrease if initially positive, or converge more rapidly if it was initially negative. This is supposedly the focusing theorem whose cause is attributed to the attractive pull of gravitation. But what about 'repulsive' gravitation in Einstein's equation? How do we know when it's there, and when it's dominant?

Under the strong energy condition, the focusing theorem is $\dot{\Theta} \leq \frac{1}{3}\Theta^2$. But all this says is that the universe is decelerating, $\ddot{S} \leq 0$. This appears to be contrary to the most recent observations. If true, however, we should have no fear of winding up in a singularity, if the concept has any meaning at all. It is rather remarkable that bulk viscosity was found to violate the energy condition. It is all the more remarkable in the sense that dissipative processes dissipate energy but do not turn sources into sinks.

2.6 Can Einstein's Equations be Considered as 'Thermal' Equations of State?

Since the Raychaudhuri equation is an identity, while the Einstein equation is an equation it may seem possible to derive the latter from the former given separate criteria that the Ricci and the energy-momentum tensors must satisfy. It didn't take too much of an imagination that the second law of a black hole might just provide such an opportunity.

The second law applied to black holes,

$$\delta Q = T dS, \tag{2.43}$$

equates the heat absorbed at the event horizon, δQ, with the entropy that is proportional to the area, A, of the event horizon. Since the left-hand side of (2.43) represents mass-energy, while the right-hand side represents geometry, it seems like the ideal setting in which to derive the Einstein equation. Moreover, the Einstein equation will appear as a thermal equation of state since the constant of proportionality in (2.43) is the absolute temperature T which we know is proportional to the surface gravity [Jac95]. This sounds almost too good to be true, and in fact, it is!

Since we are dealing with light traveling along null geodesics, the tangent vector field k^α replaces the velocity vector field, u^α, and whose components satisfy $k^\alpha k_\alpha = 0$. It looks a little fishy to see a vector normal to itself so an auxiliary null vector, N^α, is introduced to get proper normalization, $k_\alpha N^\alpha = -1$ [Wal84], where the negative sign is conventional. There appears no limit to the patchwork being done.

In place of time we have the affine parameter λ, which measures the evolution along a geodesic. The area replaces the volume, so that the expansion must be defined as the fractional rate of change of the congruence's cross-sectional area,

$$\Theta = \frac{1}{A}\frac{dA}{d\lambda}. \tag{2.44}$$

Finally, the Raychaudhuri must be modified to two space dimensions

$$\frac{d\Theta}{d\lambda} = -\frac{1}{2}\Theta^2 - \sigma^2 - R_{\alpha\beta}k^\alpha k^\beta. \tag{2.45}$$

According to (2.44), the increment in area is $\Theta A\, d\lambda$. The expansion is obtained by integrating (2.45) on the incorrect assumption that the first two terms are negligible with respect to the third. From Landau's calculation of the time component of the Ricci tensor, we know for a fact that the first and third terms are of the same magnitude. Nevertheless, following Jacobson, we integrate (2.45) disregarding the first two terms on the right-hand side to obtain $\Theta = -\lambda R_{\alpha\beta} k^\alpha k^\beta$. Thus, the expression for the increment in area is

$$dA = -R_{\alpha\beta} k^\alpha k^\beta \lambda A \, d\lambda. \qquad (2.46)$$

The heat absorbed at the horizon is related to the mass-energy tensor according to

$$\delta Q = T_{\alpha\beta} \chi^\alpha d\mathcal{V}^\beta,$$

where χ^α is a Killing vector at the horizon, and the increment in the volume is $d\mathcal{V}^\alpha = k^\alpha A \, d\lambda$. In order that the pieces, at least superficially (no pun intended), fit together, the Killing vector must be given by $\chi^\alpha = -\kappa\lambda k^\alpha$.

The surface gravity, κ, is linked to the absolute temperature via the Hawking expression, so that the second law, (2.43), where dS is strictly proportional to the area increment (2.46), now takes the form

$$\delta Q = -\kappa T_{\alpha\beta} k^\alpha k^\beta \lambda A \, d\lambda = -\text{const.} \times \kappa R_{\alpha\beta} k^\alpha k^\beta \lambda A \, d\lambda = T dS,$$

for a positive constant. This will be satisfied if

$$\text{const} \times T_{\alpha\beta} k^\alpha k^\beta = R_{\alpha\beta} k^\alpha k^\beta, \qquad (2.47)$$

for all null vectors. It implies that

$$\text{const.} \times T_{\alpha\beta} k^\alpha k^\beta = R_{\alpha\beta} k^\alpha k^\beta + f g_{\alpha\beta},$$

where the function f is fixed by the vanishing of the covariant derivative of the mass-energy tensor, which implies that $f = -R/2$ plus some arbitrary constant, like the cosmological one.

Some technical points are in order. First, on account of (2.47) and the weak energy condition, $T_{\alpha\beta} k^\alpha k^\beta \geq 0$, the increment in area, (2.46), is negative, and so too would be the increment in entropy. Jacobson got around this by defining k^α as "the tangent vector to the horozon generators for an affine parameter λ that vanishes at [point] $\mathcal{P}$ and is negative to the past of $\mathcal{P}$." This would also require a negative sign in the volume increment, $d\mathcal{V}^\alpha$, so that Einstein's equations would still have the wrong sign.

Second, what would the mass-energy tensor look like for photons? Third, we are not in $(3 + 1)$ dimensions; time has been replaced by some affine parameter, and the dimensionality of space has been reduced to two dimensions. How do we know that the Einstein equations apply to such an affine parameter-space? The vacuum solution, $T_{\alpha\beta} = 0$, about a black hole, $R_{\alpha\beta} = 0$, is necessarily flat if Einstein's equations apply. And if Einstein's equations don't apply, what is all this exercise about?

Now we turn to points of principle. We know that the Einstein equations are adiabatic so a δQ cannot be defined, let alone an entropy. By the

Einstein equivalence principle, the Hawking, or more generally the Unruh temperature, can be made to vanish. Temperatures just don't vanish on changing frames. Finally, apart from the sign, (2.46) would imply that the second law is derived from the weak energy condition, $R_{\alpha\beta}k^\alpha k^\beta \geq 0$, which it certainly does not. There is neither a relation between area and entropy nor can entropy be geometrized!

In conclusion, in general, Riemann geometry prohibits a hydrodynamic description of the universe, and in particular, a dissipative one. The Raychaudhuri identity becomes an equation for the expansion, Θ, when the density and pressure can be evaluated in terms of the it, and the stress tensor and torsion are both equal to zero.

Bibliography

[CCP65] R. H. Dicke C. Callan and P. J. E. Peebles. *Cosmology and Newtonian Mechanics*, vol. 33. 1965.

[Har00] E. Harrison. *Cosmology*. Cambridge Univ. Press, Cambridge, 2000.

[Haw66] S. W. Hawking. Perturbations of expanding universe. *Astrophys. J.*, 145:544–550, 1966.

[HE73] S. W. Hawking and G. F. Ellis. *The Large Scale Structure of Space-time*. Cambridge Univ. Press, Cambridge, 1973.

[Jac95] T. Jacobson. Thermodynamics of space-time: the Einstein equation of state. *Phys. Rev. Lett.*, 75:1260–1263, 1995.

[KS07] S. Kar and S. Sengupta. The Raychaudhuri equation: A brief review. *Pramana*, 69:49–76, 2007.

[Lav78] B. H. Lavenda. *Thermodynamics of Irreversible Processes*. MacMillan, Basingstoke, 1978.

[LL] L. D. Landau and E. M. Lifshitz. *Statistical Physics*. Pergamon Press, Oxford, 1958.

[LL51] L. D. Landau and E. M. Lifshitz. *Classical Theory of Fields*. Addison-Wesley, Reading, MA, 1951.

[LL59] L. D. Landau and E. M. Lifshitz. *Fluid Mechanics*. Pergamon Press, Oxford, 1959.

[MM34] W. H. McCrea and E. A. Milne. Newtonian universes and the curvature of space. *Quart. J. Math.*, 5:73–80, 1934.

[Pee93] P. J. E. Peebles. *Principles of Physical Cosmology*. Princeton University Press, Princeton, 1993.

[Poi04]   E. Poisson. *A Relativist's Tool Kit*. Cambridge Univ. Press, Cambridge, 2004.

[Ray79]   A. K. Raychaudhuri. *Theoretical Cosmology*. Oxford Univ. Press, Oxford, 1979.

[Wal84]   R. Wald. *General Relativity*. Chicago Univ. Press, Chicago, 1984.

[Wei72]   S. Weinberg. *Gravitation and Cosmology*. Wiley, New York, 1972.

Chapter 3

Is General Relativity Viable?

3.1 How can Mass-Energy be Conserved when Gravity is no Longer a Force?

Einstein's general relativity geometrizes the gravitational field. Since it is no longer a field of force how can it contain, no less conserve, mass-energy? As long as we are working in Euclidean space, conservation is expressed as the vanishing of a divergence. Only for Cartesian components of Euclidean space are the components of a tensor independent of the space point to which the tensor is connected to. This is no longer true for curvilinear coordinates even in Euclidean space, let alone for non-Euclidean geometries.

To say that in an isolated system, energy and momentum are conserved is valid only in Cartesian coordinates. If curvilinear coordinates are used, these quantities are no longer constant, but depend on the point they are evaluated at. It is, therefore, surprising that Einstein's theory of general relativity, which insists on a covariant formulation should use a criterion of conservation that is only valid in Cartesian coordinates. And what characterizes the gravitational field is only a 'pseudo' tensor which can be made to vanish merely by a change of coordinates! This is the culprit responsible for the inability to localize energy and momentum in space.

There is one more fundamental difference between vector spaces and manifolds. In the former we can define tangent spaces and move vectors from one tangent space to another. On manifolds this is not possible; the next best thing is to move vectors on curves. In order to do so, it is necessary

97

to introduce the notion of parallel translation which allows us to connect tangent spaces at different points on the manifold. In fact, Gaussian curvature can be thought of as the net amount that a vector turns under parallel transport around a small closed curve.

The absence of genuine conservation laws in general relativity forced Einstein to introduce, into what was supposed to be a covariant theory, a non-covariant quantity in order to obtain a conservation law. Einstein wrote down his equations:

$$G_{ik} \equiv R_{ik} - \frac{1}{2}g_{ik}R = -\kappa T_{ik}, \tag{3.1}$$

in much the same way as Newton introduced his universal law of gravitation as a postulate, even though it was motivated by Kepler's laws, and not an apple falling from a tree. The left-hand side of (3.1) is supposed to represent geometry in the form of Ricci's tensor $R_{\nu\mu}$, and its contraction R, while the right-hand side is the energy-momentum of the field, $T_{\nu\mu}$. The constant of proportionality is $\kappa = 8\pi$, in units where $G = c = 1$.

There is no valid reason why there should be an equality between the two, and its sole justification was that the vanishing of the covariant derivative of the left-hand side would impose the vanishing of the right-hand side

$$\nabla_k T_i^k = \frac{1}{\sqrt{-g}}\frac{\partial(T_i^k\sqrt{-g})}{\partial x^k} - \frac{1}{2}\frac{\partial g_{kl}}{\partial x^i}T^{kl} = 0. \tag{3.2}$$

According to Landau and Lifshitz [LL75] , and just about everyone else, "this equation does not express any conservation law whatever." This is "because the integral $\int T_i^k\sqrt{-g}\,dS$ is conserved only if the condition

$$\frac{\partial T_i^k\sqrt{-g}}{\partial x^k} = 0 \tag{3.3}$$

is fulfilled, and not (3.2)." Møller [lle58] comes out and says that if the second term in (3.2) is different from zero, "then it expresses the fact that matter energy is not conserved." Landau and Lifshitz claim further that the four momentum of the matter must not be conserved alone, but the combined matter and gravitational field. This, however, is contested by the fact that tensor divergence of the Einstein tensor G_{ik} on the left-hand side of (3.1) is identically zero. So if the vanishing of the tensor divergence on the left-hand has meaning, as we know it does, so too must the vanishing of the tensor divergence on the right-hand side, namely (3.2).

3.2 Gauss' Theorem and Einstein's Condition of Conservation

Condition (3.3) is what Landau and Lifshitz claim as the condition for conservation in their footnote to (96.1). Yet, going back to (86.14), they give Gauss' theorem as

$$\int \nabla_i X^i \sqrt{-g} \, d\Omega = \oint X^i \sqrt{-g} \, dS_i$$

where Ω is a four-volume bounded by the closed surface, S. But what about a Gauss theorem for non scalar quantities?

Technically speaking, a volume integral of a vector over some finite volume cannot be a vector because the laws governing a vector are different at different points. For a tensor of second rank, Folomenshkin [Fol72] proposes

$$\int \nabla_k (T_i^k X^i) \sqrt{-g} \, d\Omega = \oint T_i^k X^i \sqrt{-g} \, dS_k,$$

for any arbitrary vector, X^i.

Conservation equations of the form (3.3) are invariant with respect to linear coordinate transformations. Einstein was considering such types of invariant equations during the period 1913–1914. According to Folomeshkin, "Einstein non-critically carried over equations (3.3) into the newly general covariant theory. As a result, the 'problem' of energy-momentum arose."

Undoubtedly, the simplest way to patch things up was to consider

$$\frac{\partial \sqrt{-g} \, t_i^k}{\partial x^k} = -\sqrt{-g} \, \frac{1}{2} \frac{\partial g_{lm}}{\partial x^i} T^{lm},$$

by introducing some non tensorial quantity, t_i^k, and writing the non-covariant conservation law as

$$\frac{\partial}{\partial x^k} \left[\sqrt{-g} \left(T_i^k + t_i^k \right) \right] = 0. \tag{3.4}$$

Einstein had this to say about the second term in (3.4):

> From a physical point of view, the presence of the second term on the left-hand side means that for matter alone the conservation laws of momentum and energy are not satisfied in a real sense; more precisely, they are satisfied only when the g_{ik} are constant, i.e., when the components of the gravitational field intensity are zero. This second term is the expression for the momentum and, accordingly, the energy that in a unit time and in a unit volume are transferred from matter to the gravitational field.

Since (3.4) is not a covariant conservation equation, it would essentially allow energy to be created out of nothing!

The vanishing of the covariant derivative does lead to a conservation law, but not one we Euclideans are use to dealing with. In fact, we will now see that it can account for the bending of light by a massive body and the advance of the perihelion through a coupling of angular momentum and gravity.

With this end in mind, consider a geodesic parametrized by arc length λ, and denote $\dot{x}^i = dx^i/d\lambda$. The covariant derivative of a contravariant vector X^k can be written as

$$\nabla_i X^k \; \dot{x}^i = \frac{\partial X^k}{\partial x^i} \dot{x}^i + \Gamma^k_{ij} X^i \dot{x}^j = \frac{dX^i}{d\lambda} + \Gamma^k_{ij} X^i \dot{x}^j.$$

For geodesics, the covariant derivative vanishes, and setting $X^i = \dot{x}^i$ there results in

$$\ddot{x}^k + \Gamma^k_{ij} \dot{x}^i \dot{x}^j = 0. \tag{3.5}$$

This is the equation of motion for a geodesic.

3.3 Non-conservation of Angular Momentum: Bending of Light and Advance of the Perihelion

Consider the Beltrami metric,

$$ds^2 = \frac{dr^2}{(1 - r^2/R^2)^2} + \frac{r^2 d\varphi^2}{1 - r^2/R^2}, \tag{3.6}$$

in polar coordinates r, φ, with R as the absolute constant. For the Beltrami metric (3.6), the Levi-Civita connection is

$$\Gamma^2_{12} = \frac{2}{r(1 - r^2/R^2)},$$

so that the associated geodesic equation is

$$\ddot{r} + \frac{2}{r(1 - r^2/R^2)} \dot{r}\dot{\varphi} = 0.$$

Integrating, we get

$$\frac{r^2 d\varphi/d\lambda}{1 - r^2/R^2} = J. \tag{3.7}$$

If it were not for the denominator differing from unity, (3.7) would be the law of the conservation of angular momentum per unit mass in polar coordinates, with λ representing coordinate time — not proper time for which the denominator is again unity. However, in view of the time-independent Beltrami metric we have no call to make such a distinction.

If we switch from constant density to constant mass by introducing the Kepler (escape!) velocity law, required for the Schwarzschild metric,

$$\frac{r^2}{R^2} = \omega^2 r^2 = \frac{2M}{r}, \tag{3.8}$$

where the last term is the ratio of the Schwarzschild radius to r — that is characteristic of all relativistic corrections — then (3.7) becomes the same expression for the angular momentum per unit mass found in the Schwarzschild solution,

$$\frac{r^2 \dot{\varphi}}{1 - 2\alpha/r} = J, \tag{3.9}$$

where $\alpha = M$ is an 'arbitrary' constant of integration.

Analogously, for the linear momentum per unit mass, the equation for the geodesic (3.5) becomes

$$\ddot{r} + \Gamma_{11}^1 \dot{r}\dot{r} = 0, \tag{3.10}$$

where the connection is

$$\Gamma_{11}^1 = -\frac{4\alpha/r^2}{1 - 2\alpha/r}.$$

Integrating (3.10), we find the covariant conservation of the linear momentum, per unit mass, as

$$\frac{\dot{r}}{(1 - 2\alpha/r)^2} = P,$$

which happens to be the exact same expression found using the Schwarzschild metric.

In fact, the bending of light and the advance of the perihelion are to be found in the covariant conservation of the angular momentum, (3.9). Writing $\ell = r^2 \dot{\varphi}$, the centrifugal energy is

$$\frac{\ell^2}{2r^2} = \frac{J^2}{2r^2}\left(1 - \frac{4\alpha}{r}\right),$$

where terms of first order only in the Schwarzschild radius have been kept. When this term is added onto the gravitational potential we get what is referred to as the Einstein potential,

$$V_E = -\frac{GM}{r} + \frac{J^2}{2r^2}\left(1 - \frac{4\alpha}{r}\right),$$

in contrast to the Newtonian potential where the last term is absent.

It is ℓ, the angular momentum per unit mass, which is constant. Note also that Kepler's law would have sufficed in (3.8) as far as the numerical result for the bending of light by a massive body is concerned. This is the famous doubling of Einstein's estimation of the bending of light which he obtained by treating a variable speed of light, in blatant contradiction to his special theory of relativity.

The apparent non-conservation of J, or P, is not due to any dynamical effect, but, rather, to the non-Euclidean nature of the geometry so that $\nabla_i J^i = 0$ does not imply that $J^i(r) = $ const. This can also occur in Euclidean geometry if we use curvilinear coordinates. So it is, in general, the vanishing of the covariant derivative, and not $\partial J^i/\partial x^i = 0$, which determines energy-momentum conservation. Hence, Einstein's condition of conservation does not hold water.

This has been emphasized in the past by Folomeshkin [Fol72], but seeing the number of citations it received, it has fallen on deaf ears. The problem is parallel translation of the vector $J^i(r)$ from r to another point r'. In general, we will not get the vector $J^i(r')$, but rather, a new vector $J^{i\,\prime}(r')$.

Even if we accept (3.2) as a covariant statement of conservation, what is being conserved? The second term in (3.2) depends upon a varying gravitational field so matter is receiving energy and momentum from the gravitational field. It does not even make sense to talk about an isolated system upon which to tally the books of energy additions and withdrawals since gravity is an infinite range force.

However, gravity is not a physical field in Einstein's general relativity, where it has been geometrized. Then what is the role of his equivalence principle which can annihilate the effects of gravity locally? How do you iron out the curvature of non-Euclidean geometry through a coordinate transformation?

Einstein supposed that the energy density of the gravitational field could be made zero in contrast to those of the electromagnetic field. In a paper dealing with gravitational waves, Einstein wrote "It could be that by an appropriate choice of coordinates, one could always make the components of the energy of the gravitational field vanish, which would be extremely interesting." Luckily, he never achieved his goal.

3.4 When Einstein meets Newton

Geometry is buried in the values of the metric tensor, g_{ik}. Even if Einstein's equations (3.1) were beyond reproach, it would be necessary that the spatial metric reduce to the well-known forms of non-Euclidean geometries in the constant curvature limit, as, for example, in the case of the Beltrami metric, (3.6).

If Gauss' theorem is to be used where do we take the surface of integration? It seems natural to use the surface of integration in the asymptotically flat region as $r \to \infty$. This desire appears to be due to Einstein who, while outright rejecting absolute space, clings onto *ferne massen* (distant masses) in order that the gravitational field should be zero at infinity. Only a theory of absolute space (and time), like Newton's, prescribes the constant of integration, while in Einstein's theory the differential equation is prescribed and the constants of integration are relative.

In de Sitter's [dS17] words: "The condition that the gravitational field shall be zero at infinity forms part of the conception of an absolute space, and in a theory of relativity it has no foundation." de Sitter had higher hopes for Einstein's general theory of relativity than what actually materialized.

This still leaves us with the inability to localize energy and momentum in space, and why such a localization should require [lle58] "the use of quasi-Galilean coordinates in which the components of the metric tensor, for increasing spatial distances, sufficiently rapidly converge towards the constant values of the special theory of relativity." But Møller should have been aware that in the weak gravitational limit, gravity has become a physical field, so that it is difficult to believe that what was has been geometrized, has simply been transformed into a field of force merely by going to the asymptotic limit!

Where is the real seat of difficulty with Einstein's theory? By setting the pseudo tensor equal to the extra piece in the definition of the covariant derivative we obtain a covariant conservation condition for the energy-matter tensor. But, it does not mean that the energy-matter tensor is a constant since it depends on some nonlocal translation law from some point on the surface S to some common reference point, which may be the center of mass. In other words, the sum of two tensors has no sense without the specification of the point in the curvilinear system to which we specify the sum of the tensors [Fol72].

General relativity is rather cavalier in eliminating divergences. The Lagrangian that Einstein started with was $\sqrt{-g}\,R$. It contains both the first and second derivatives of the metric tensor. However, Einstein observed that all the second derivatives were contained in the total derivative of a certain quantity, ω^i, whose divergence differs from Ricci's scalar by a non-covariant density $\sqrt{-g}\,G$,

$$\frac{\partial \omega^i}{\partial x^i} = \sqrt{-g}(R - G). \qquad (3.11)$$

G is the difference of products of the connections Γ_{ij}^k, which in turn, are sums of first derivatives of the g_{ik}, and hence, is not a tensor. It is the actual quantity that Einstein used to derive his pseudo tensor density, $t_i^k \sqrt{-g}$. And since it too contains only the metric components and their first derivatives, it likewise doesn't qualify for an invariant quantity, which must contain second derivatives that measure the bending of the surface, as in Gauss' expression for the curvature of a surface.

Ordinarily, the volume integral of a divergence can be transformed into a surface integral that vanishes in the asymptotic limit as $r \to \infty$. As Logunov and Folomeshkin [LF97] point out this is true when the field variables decrease as $1/r$. However, the field variables in Einstein's theory, g_{ik}, do not tend to zero as $r \to \infty$ so that the surface integrals do not vanish in the asymptotic limit of flat spacetime. Recall our earlier discussion where general relativity makes no restrictions on the g_{ik}, which can even be infinite at infinity! Einstein presented de Sitter with one such example.

Since the divergence term in (3.11) can't be gotten rid of, and although it does not alter the field equations, it nevertheless changes the physical content of the theory. For a non-static metric, the field equations show that $\partial \omega^0 / \partial x^0$ contributes to the expression for the mass [Fol74].

Fock [Foc59] points out that if you give up asymptotic uniformity and flatness, you will undoubtedly get a violation of some, if not all, of the conservation laws. This is because the properties of uniformity and flatness express the fact that the system is isolated.

Consider the Schwarzschild solution in this respect. Schwarzschild introduced the metric,

$$ds^2 = e^{\nu(r)} dt^2 - e^{\lambda(r)} dr^2 - r^2(d\varphi^2 + \sin^2 \varphi \, d\vartheta), \qquad (3.12)$$

with $\nu(r)$ and $\lambda(r)$, as unknowns, into Einstein's equations, (3.1), and sought a static solution in which the two unknowns were independent of time, but could depend on the radial coordinate, r. He, moreover, required that the solution be in empty spacetime, and used Einstein's condition of emptiness,

$$G_i^k = 0. \tag{3.13}$$

He then found

$$e^{-\lambda(r)} = 1 - \frac{2\alpha}{r}, \tag{3.14}$$

where α is independent of r, but on matching it to the Newtonian limit found M for its value. Going to the full asymptotic limit of free space, where the metric tensor becomes independent of r, we regain special relativity. However, as we have mentioned, special and general relativities are incompatible with one another so the boundary condition is illusionary.

Riemann curvature, R_{ijkl}, is intrinsic curvature since it is a function of the connections, Γ_{ij}^k, which are intrinsic. The Ricci tensor, R_{ij}, is a kind of a trace of the Riemann curvature. It can be interpreted as a sectional curvature. Again taking the trace we come to the scalar curvature, R. Now comes the punch line: the Riemann tensor cannot be proportional to the density of matter in any way; otherwise, the Riemann tensor would vanish in empty space. Quoting Robertson and Noonan [RN68], the vanishing of the Riemann tensor in empty space "would imply that gravity does not act in empty space, contrary to observation." No mass no gravitational attraction, and consequently, no gravity!

In the 'quasi' asymptotic limit, where the metric tensor components differ little from unity, Newton's law must be respected, and in this weak field approximation, $R_0^0 = -\nabla^2 \Phi$, where Φ is the Newtonian gravitational potential. Comparing this with Poisson's equation, $\nabla^2 \Phi = 4\pi\varrho$, associates the time component of the Ricci tensor with the mass density, $R_0^0 = -4\pi\varrho$. The metric component is $g_0^0 = 1 + 2\Phi$ to lowest order in the gravitational potential. So, it would appear that just by taking the trace of the Riemann tensor, which in no way depends on the density of matter, we get the Ricci tensor whose time component is proportional to it!

Even if we were to accept that the asymptotic weak field limit should coincide with the Newtonian dynamics, Einstein's field equation,

$$R_0^0 - \frac{1}{2}Rg_0^0 = -8\pi T_0^0,$$

will not be satisfied because the scalar curvature, R, cannot be determined in terms of the gravitational potential, Φ, given the same accuracy that was used to determine R_{00}. Although there are artifices for getting around this, it does blemish a theory that should be intrinsically coherent.

By the same line of (mis-)reasoning, the constant in (3.14), α, is set equal to the mass, M. The mass of what though? — seeing that the Einstein's equations were solved under Einstein's condition of emptiness, (3.13). The mathematician, O'Neill [O'N83] tells us that we shouldn't be concerned because we are not 'modeling' the mass, M.

In fact, mathematicians have become the standard-bearers of general relativity by throwing their unwavering — and uncritical — support behind it, beginning with Hilbert who was vying with Einstein to be its sole author. Most surprisingly, Hilbert's mentor Felix Klein, the most gifted geometer of the 19th century, even wrote articles on general relativity. Undoubtedly, general relativity has given their field the physical credence that it would otherwise not have had. So what is being modeled? Since mathematicians don't worry about such particulars as physics, it seems perfectly alright not to model M.

Rather, if we want to 'model' the mass, the Einstein equation for the time component would be

$$G_0^0 = -\frac{1}{r^2}\frac{d}{dr}\left[r\left(1 - e^{-\lambda(r)}\right)\right] = -8\pi\, T_0^0.$$

Integration gives the mass as

$$M = \int_0^{r_0} T_0^0\, 4\pi r^2 dr = -\frac{1}{2}\int_0^{r_0}\frac{d}{dr}[r(1 - e^{-\lambda(r)})]dr = -\frac{1}{2}r(1 - e^{-\lambda(r)})\Big|_0^{r_0}.$$
(3.15)

Therefore, when T_0^0 vanishes so, too, will M, thereby validating Einstein's condition of emptiness, (3.13). The interval of integration in (3.15) can hardly be considered as asymptotically remote from all masses that would justify the weak field limit where Newtonian gravity would be valid. Furthermore, $e^{-\lambda(r)} = 1 - 2Mr_0^{n-1}/r^n$, for any $n > 1$, would work just as well as (3.14), as far as (3.15) is concerned. Recall that integration constants are relative in general relativity!

And since we are supposedly in the Newtonian limit, we can use the Newtonian result that a uniform sphere exerts the same attraction as a point mass at its center (i.e., Newton's shell theorem). This being the case, what is the significance of the vanishing of (3.14) at $r = 2M$?

Why did Schwarzschild look for a static solution to a spherical symmetrical system in the form of (3.12), and not in the form

$$ds^2 = e^{\nu(r)}dt^2 - e^{\lambda(r)}(dr^2 + r^2 d\vartheta^2 + r^2\sin^2\vartheta d\varphi^2)?$$

The asymptotic solution, as $r \to \infty$, is

$$e^{\lambda(r)} = \left(1 + \frac{M}{2r}\right)^4 ,$$

and therefore, in contrast with (3.14), there is not even a smidgen of a singularity at $r = 2M$. Much more is left to chance in general relativity than to design!

3.5 The Lack of a Gravitational Invariant

As we have already seen, the variation of the non covariant Lagrangian, G, determines the pseudo tensor density, $\sqrt{-g}\ t_i^k$, which is a function of g_{ik}, and their first derivatives. As such it is not a differential invariant, which requires second derivatives because first derivatives do not determine curvature. Moreover, it can be made equal to zero locally at a point.

Even more can be said in the case of a plane gravitational wave [LF97]. For in this case, the derivatives of the metric tensor can be made to vanish on the entire surface that is normal to the direction of propagation of the wave. This leads to the conclusions that there are no gravitational waves in Einstein's general relativity, and even more important, energy and matter cannot be localized.

There are those who consider the lack of a privileged system of coordinates, and a lack of a precise definition of what the coordinates stand for, real virtues of general relativity. This is what de Sitter has emphasized. However, Fock would ask him are we at the same liberty to choose between the heliocentric theory of Copernicus and the geocentric theory of Ptolemy?

Why should special relativity be the asymptotic limit of general relativity? The former treats uniform motion, which in the presence of gravity no longer exists. What is relative in the general theory? The special theory treats gravity as a force, but the general theory tries to geometrize it. How can dynamics be obtained from pure geometry? And how can the two concepts be harmonized?

There is not one test of general relativity that cannot be gotten by other less drastic measures, and by much more intuitive means. For these tests, spacetime metric splits into two pieces: the gravitational red shift which is treated by the coefficient of the square of the time interval, and the static tests of the deflection of light and the advance of the perihelion which deal with the radial component of the metric, assuming spherical symmetry.

Then on what basis is it assumed that gravity propagates at the speed of light? If that were true the planets could not maintain their orbits about the sun, because delayed action would result in a torque on planetary motion, and gravity would manifest aberration, just like light does. That these effects do not occur, and the general theory of relativity does not tell us what gravity is, leave much to be desired.

Last but not least, there's the psychological issue of why Einstein's theory is so idolized. Even Einstein felt uneasy about the predictions of his theory. It is well-known that he was staunchly against the singularities found in his theory, and the fact that he obtained a positive energy flux for gravitational waves. This was in contradiction to $\sqrt{-g}t_i^i = G$ not being an invariant, which is enough to render it inadmissible as a true density of gravitational radiation [LC17]. As a consequence, spontaneous waves could give rise to energy dispersion through radiation. For, according to Einstein, "Since this fact should not happen in nature, it seems likely that quantum theory should intervene by modifying not only Maxwell's electrodynamics, but also the new theory of gravitation." It does appear very odd indeed that Einstein should make an appeal to a theory he thought to be incomplete, and one he did not believe in.

Bibliography

[dS17]   W. de Sitter. On the relativity of rotation in Einstein's theory. *KNAW Proceedings*, 19:527–532, 1917.

[Foc59] V. Fock. *The Theory of Space, Time and Gravitation.* Pergamon Press, Oxford, 1959.

[Fol72] V. N. Folomeshkin. Covariant integration of tensors, energy-momentum in general relativity. Technical report. IHEP 72-92, 1972.

[Fol74] V. N. Folomeshkin. Zero value of the Schwarzschildian mass of asymptotically Euclidean time symmetrical gravity waves. *International Journal of theoretical Physics*, 10:145–151, 1974.

[LC17]   T. Levi-Civita. On the analytic expression that must be given to the gravitational tensor in Einstein's theory. *Rendiconti della Reale Accademia dei Lincei*, 26:381, 1917.

[LF97]   A. A. Logunov and V. N. Folomeshkin. Energy-momentum of gravitational waves in the general theory of relativity. *Theoretical and Mathematical Physics*, 32:667–672, August 1997.

[LL75]   L. D. Landau and E. M. Lifshitz. *Classical Theory of Fields.* Pergamon Press, Oxford, 1975.

[lle58] C. Møller. On the localization of the energy of a physical system in the general theory of relativity. *Annals of Physics*, 4:347–371, 1958.

[O'N83] B. O'Neill. *Semi-Riemann Geometry with Applications to Relativity.* Academic Press, San Diego, 1983.

[RN68]   H. P. Robertson and T. W. Noonan. *Relativity and Cosmology.* W. B. Saunders, Philadelphia, 1968.

Chapter 4

Is Gravity an Emergent Phenomenon?

4.1 Thermodynamics Meets Cosmology

Thermodynamics has met cosmology at the horizon thanks to a paper, discussed in §2.6, by Jacobson who attempted — unsuccessfully however — to derive the Einstein equations as 'thermal' equations of state from the 'second law' for black holes.

Emergent gravity, a term coined by Sakharov in 1967, implies that gravity is not a fundamental force of nature, but one that can be accounted for by other, more fundamental, laws of nature. The putative transmitter of the gravitational force, the graviton, would be 'demoted' from its elementary particle status, as that of the photon, to a mere excitation, like a phonon in a crystal [DAESb].

First, and foremost, statistical principles cannot govern deterministic laws, and neither can they derive them. Second, having entropy at the horizon creates more problems than it attempts to solve. Third, energy, whether at the horizon or anywhere else is in conflict with the adiabatic nature of Einstein's field equations. And finally, if gravity is an emergent phenomenon it is certainly taking its sweet time to emerge, as we shall now show.

4.2 Negative Pressure or Positive Tension?

Although there is massive confusion in the literature, tension is not negative pressure, and radial distance is not analogous to volume. *In unconstrained thermodynamic systems, entropy increases with with volume thereby defining (positive) pressure, while in constrained thermodynamic systems, entropy decreases with elongation, or stretching, thereby defining (negative) tension.*

Entropy, S, increases with volume,

$$\left(\frac{\partial S}{\partial V}\right)_E = \frac{p}{T} \geq 0, \tag{4.1}$$

at constant internal energy E, where p is the pressure, and T, the absolute temperature. Unlike the former, the latter is always positive semi-definite. Definition (4.1) follows from the differential of Euler's relation,

$$dS = \frac{1}{T}\left(dE + p\, dV\right), \tag{4.2}$$

for a first order homogeneous function.

If the pressure were to become negative, the body would have to spontaneously contract so as to increase its entropy. States with negative pressure are only metastable, since spontaneous contraction involves the formation of cavities having smaller volumes than the original one [LL58].

Examples of constrained thermodynamic systems are those which are stretched, squeezed, or confined in some other way. Their entropies undergo a reduction, $S(r)$, as a result of such actions from the maximal possible entropy, $S(0)$, in the unconstrained state [Lav95].

To derive an expression for the entropy reduction, it suffices to consider a symmetric random walk, whose probability density for a jump of dr is

$$f(r)\, dr = \frac{1}{\sqrt{2\pi \langle r^2 \rangle}} \exp\left(-\frac{r^2}{2\langle r^2\rangle}\right) dr,$$

where $\langle r^2 \rangle$ represents the average square end-to-end distance of the random walk. The entropy lost through the elongation is

$$S(r) - S(0) = -\ln f(r) = -\frac{r^2}{2\langle r^2 \rangle} + \text{const.}, \tag{4.3}$$

since entropy is defined classically to within a constant factor. The tension causing the stretching is

$$\tau/T = -\frac{dS}{dr} = r/\langle r^2 \rangle, \tag{4.4}$$

is a Hookean law. This is a consequence of the gaussianity of the process, i.e., small deviations from the unconstrained state of maximum probability.

Now, the big question is: Which pressure does Einstein's equations use to get an expanding universe, the negative of (4.1) or (4.4)? Obviously, the negative of (4.1) because Einstein's equations satisfy (4.2) when it is equal to zero, i.e., an adiabatic process.

4.3 Are Deterministic Laws Governed by Statistical Principles?

Verlinde [E.V] argues that gravity is an entropic force near the horizon of a black hole, which serves as a heat bath. He then asks us to consider a particle located about a Compton's wavelength from the horizon of the black hole. We can think of the horizon as a holographic screen which can be used to store information, whatever that means.

He assumes that the displacement, Δx, will cause a change in the entropy of the screen by an amount,

$$\Delta S = 2\pi k \frac{\Delta x}{\lambda_C}, \tag{4.5}$$

where $\lambda_C = \hbar/mc$ is the Compton wavelength of a particle of mass m. Displacements of single particles do not, however, give rise to net entropy changes!

Nonetheless, the change in entropy will give rise to a force, F, defined by

$$F/T = \Delta S/\Delta x, \tag{4.6}$$

which looks a lot like (4.4). Since the right-hand side is fixed by (4.3), we can either choose the force or the temperature, which will then determine the other. Velinde chooses the latter, and sets it equal to the Unruh temperature,

$$T = \frac{\hbar a}{2\pi k c}, \tag{4.7}$$

and lo and behold he comes out with Newton's second law, $F = ma$.

However, Verlinde already sees a fly in the ointment because the constant c has appeared in what amounts to a non relativistic derivation of Newton's law. So, he asks us to scrap the derivation, and to start all over again.

We are now asked to consider the relation between the number of bits (of what?) and the area A needed to store the bits on the screen,

$$N = A/\ell_{Pl}^2. \tag{4.8}$$

This begs the question for it is a camouflaged form of the Bekenstein expression for the entropy of a black hole, where $\ell_{Pl} = \sqrt{G\hbar/c^3}$ is the Planck length. The number of bits is (just!) the number of degrees of freedom in the law of equipartition of energy,

$$E = \frac{1}{2}NkT, \tag{4.9}$$

of an ideal gas. Black holes, information, Newton's law, the speed of light, and a classical ideal gas form a perfect goulash.

Finally, with $E = Mc^2$ as the rest energy, Verlinde comes out with the Newtonian gravitational law, $F = GMm/r^2$, upon introducing the expression for the area, $A = 4\pi r^2$. But, why stop here? Introducing the Newtonian gravitational law into (4.6), and then into (4.3) result in the distance of separation of the two masses by what amounts to the geometric mean of the classical Schwarzschild radius, GM/c^2, and the ultrarelativistic thermal wavelength, $\lambda_T = \hbar c/kT$. How Newton's law of gravitation can come out of such a concoction of unrelated equations is beyond any rationale whatever. It is, however, a display of maximum ignorance.

The motion of a single particle generates as much entropy change as a monochromatic beam of radiation. Lorentz was well aware of this when he was determining the frequency dependency of the entropy of black body radiation. In order to consider entropy, it was necessary to consider *bundles* of rays. Single particle dynamics does not warrant a statistical description of great number of identical particles. There is absolutely no relation between (4.4) and (4.6). But, this should not discourage us from continuing for there are more absurdities to uncover.

Verlinde later admits that (4.9) is not equipartition of energy, since $N = 4S/k$ if (4.8) is to reproduce the exact form of the Bekenstein-Hawking expression of the entropy. Then, (4.8) is really $E = 2TS = 2SdE/dS$, which integrates to the quadratic expression of the entropy as a function E, what ever E happens to be. If (4.9) were really the classical law of equipartition, N, would be a constant, and not a quadratic function of the energy. Apart from the fact that entropy cannot be a function of a power of the energy greater than unity, what is the sense of confusing the reader with completely extraneous relations?

Not only are concepts confused, but also are vectors and scalars! Associating a temperature with the gradient of the Newtonian potential, Φ, i.e., $kT = m\lambda_C \nabla\Phi/2\pi$, is bad enough, but to go further and specify that "the derivative is taken in the direction of the outward pointing normal to the screen" makes things only worse. Why this?

Because Verlinde wants to deduce the Unruh expression, where T is the temperature of the black body radiation that the accelerating detector registers to a static measurement of temperature in the gravitational field of Φ, and in so doing derive Poisson's equation. Since the radiation is that of a black body, the acceleration is fixed at $a = c^2/\lambda$, which reduces it to Wien's displacement law, $\lambda T = \text{const}$. What Wien's displacement law has to do with an accelerating observer is beyond comprehension. This only adds more ingredients to the goulash.

The law of equipartition, (4.9), is now written in the integral form:

$$E = \frac{1}{2}k \int_{\mathcal{S}} T \, dN, \qquad (4.10)$$

where the integral is taken over the screen $\mathcal{S}$. Why would we ever want to integrate a law of equipartition? Replacing left-hand side of (4.10) by Mc^2, and the right-hand side by $dN = c^3 dA/G\hbar$ give

$$M = \frac{1}{4\pi G} \int_{S} \nabla\Phi \cdot d\mathbf{A}. \qquad (4.11)$$

Where the unit normal vector, $\mathbf{n}$, in $\mathbf{A} = \mathbf{n}A$, came in is left as a deep unsolved mystery.

Anyhow, it allows Verlinde to appeal to Gauss' law to convert the surface integral (4.11) into a volume integral so that he arrives at

$$M = \int_{V} \varrho \, dV = \frac{1}{4\pi G} \int_{V} \nabla^2\Phi \, dV,$$

which is, formally if nothing else, Poisson's equation when the integrands are equated.

There are, however, more problems: T is not a function of N in (4.10), and screen's surface, $\mathcal{S}$, doesn't enclose a volume V. If this is the emergent property of gravity, it will take an awfully long time to emerge!

The sloppy reasoning makes itself manifestly apparent, and not worthy of a freshman's math errors. But, what is not so apparent is the extensive media coverage that it drew, and the slew of papers that found there way into arXiv, and some respectable scientific journals. There is no excuse for the *New York Times* to publish an article with the impressive title,

"A scientist takes on gravity" with the ridiculous assertions that "gravity is a consequence of the venerable laws of thermodynamics which describe the behavior of heat and gases," and "gravity is simply a byproduct of nature's propensity to maximize disorder." This is simply a display of ignorance. Perhaps the columnists for the *New York Times* can be pardoned but not the 'experts' that the computer technicians rely on at arXiv.

4.4 Entropic Everything

The physics community is indeed very susceptible to colds: when one person sneezes the rest catch it. There have been many variations on same theme: Begin with the Unruh temperature, (4.7), introduce the acceleration, $a = GM/r^2$, evaluated at the Schwarzschild radius, $a = c^4/4GM$, and out comes the Hawking temperature, $T = \hbar c^3/8\pi kGM$. Setting the acceleration, $a = cH$, where H is the Hubble constant, in (4.7) gives a temperature,

$$T_H = \frac{\hbar H}{2\pi k} \sim 3 \times 10^{-30} K. \tag{4.12}$$

Since the Hubble constant is proportional to the velocity at which the universe expands, we have a temperature, (4.12), that is now proportional to the velocity, and would be in contradiction with an Unruh temperature since we have no way of knowing that an acceleration exists.

This is reminiscent of Einstein's [Ein11] first derivation of the frequency shift due to the deflection of light by a massive body. He begins with the classical Doppler shift. Then he writes the constant velocity as the product of gravitational acceleration and time. For the time he substitutes the ratio of the distance and the velocity of light. Now he substitutes the ratio of the velocity to the velocity of light by the ratio of the gravitational potential — gravitational acceleration times distance — and the square of the speed of light. What was a measurement at constant velocity has miraculously been transformed into a static one!

The Unruh temperature contains both Planck's constant and the speed of light. We know, therefore, that it is quantum in nature and ultrarelativistic. Multiplying numerator and denominator by the Compton wavelength, $\lambda_C = \hbar/mc$, of a particle of mass m gives $T = \Phi m/2\pi k$, where — like Einstein — we have substituted the gravitational potential Φ for $a\lambda$. Now, the Unruh temperature is entirely classical. This is the dimensionality game all over again, and it appears very much like pulling rabbits out of a hat.

But 'No' say Easson, Frampton and Smoot [DAESa] (EFS) because (4.12) combines with the acceleration of the horizon in the way provided for by (4.7),

$$a_{hor} = 2\pi ckT/\hbar = cH \sim 10^{-9}m/s^2. \tag{4.13}$$

According to EFS, (4.13), when (4.12) is used to evaluate it, gives "a cosmic acceleration essentially in agreement with observation." This is not surprising at all since the second equality is a mere identity. What is surprising though is the smallness of the horizon temperature, (4.12), which EFS claim replaces the zero of temperature in the third law. Nothing remains sacred in the advancement of science, but whose science?

EFS invokes the entropic force to calculate the negative pressure required by them for expansion. The entropy on the Hubble horizon $R_H = c/H$ is

$$S_H = \frac{kc^3}{G\hbar}\frac{A}{4} = \frac{kc^3}{G\hbar}\pi R_H^2. \tag{4.14}$$

Now increasing the Hubble radius R_H by an amount Δr, which presupposes decreasing the Hubble constant, the increase in entropy will be

$$\Delta S_H = \frac{kc^3}{G\hbar}2\pi R_H \Delta r.$$

The entropic force is 'simply':

$$F = -\frac{dE}{dr} = -T\frac{dS}{dr}, \tag{4.15}$$

"where the minus sign indicates pointing in the direction of increasing entropy or the screen, which in this case is the horizon." A simple cold has turned into influenza!

Continuing we read:

The pressure from the entropic force exerted is

$$p = \frac{F}{A} = -\frac{T}{A}\frac{dS}{dr} = -\frac{c^2H^2}{4\pi G} = -\frac{2}{3}\rho_c c^2, \tag{4.16}$$

where $\rho_c \equiv 3H^2/8\pi G$. This is close to the value of the currently measured dark energy/cosmological constant negative pressure (tension). In this case the tension does not arrive from the negative pressure of dark energy but from the entropic tension due to the entropy content of the horizon surface.

So the entropy is increasing outwards, and this is due to a negative pressure, (4.16). Entropy cannot be localized on a surface but, rather, is a property of the volume. Therefore, in place of (4.14), EFS should have written

$$S = \left(\frac{3}{4\pi}\right)^{2/3}\frac{kc^3}{G\hbar}\pi V^{2/3}.\qquad(4.17)$$

The pressure is simply

$$\left(\frac{\partial S}{\partial V}\right)_E = \frac{p}{T} = \frac{1}{2}\frac{kc^3}{G\hbar}\frac{1}{R_H}.\qquad(4.18)$$

And evaluating the temperature with (4.12) results in a positive pressure,

$$p = \frac{c^2 H^2}{4\pi G} = \frac{2}{3}\rho_c c^2,\qquad(4.19)$$

which is exactly what we would expect for an entropy increasing with volume [c.f., (4.1)].

It is clear that (4.4) is not (4.15); in the former the entropy is decreasing with increasing constraint whereas in the latter the entropy is increasing as the radius increases. Moreover, from (4.18) we have the thermodynamic stability condition that the isothermal compressibility, $-(\partial V/\partial p)_T/V$ is positive, while from (4.15) we find

$$\frac{dS}{dr} = \frac{kc^3}{G\hbar}2\pi R_H > 0.$$

To have tension, the entropy must decrease with stretching, as (4.4) plainly shows. The force, (4.15), consequently, is *not* pointing in the direction of increasing entropy.

4.5   Derivation of the Friedmann-Lemaître Equations from the First and Second Laws

In an Appendix to their paper, EFS [DAESa] 'derive' the Friedmann equations from what they refer to as the first law. Actually, it is a combination of what formally look like the first and second laws. It is worth repeating the derivation if only because it underscores the massive confusion between the internal energy density and mass density. Since a perfect fluid is being considered, the latter is constant. For simplicity, we will consider the case of vanishing curvature where the 'apparent' horizon is the same as the Hubble horizon.

The energy flow across the horizon is

$$-dE = c^2 dM + p dV_H = (\rho c^2 + p) dV_H = (\rho c^2 + p) A_H \dot{R}_H dt$$
$$= (\rho c^2 + p) A_H H R_H dt, \tag{4.20}$$

where V_H is the Hubble volume whose surface area is A_H, and Hubble's relation, $\dot{R}_H = H R_H$ has been used. But, is this Hubble's relation? Because on differentiating $R_H = c/H$, with respect to time, we get $c^2 \dot{R} = -R_H^3 H \dot{H}$, and so do EFS. Hence, $\dot{R}_H$ has been confused with the recession velocity v in Hubble's law.

Where this energy flow comes from is anyone's guess. In any case, it implies there is something beyond the horizon. The Hubble horizon has an entropy, (4.14), and temperature, (4.12), so the first law (actually it is the second when the sign is changed!):

$$-dE = (\rho + p/c^2) c^2 A_H H R_H dt = T dS_H = \frac{H c^3}{8\pi G} dA_H = \frac{H c^3}{2G} R_H^2 dR_H. \tag{4.21}$$

Now, the equation,

$$\dot{\rho} + 3H(\rho + p/c^2) = 0, \tag{4.22}$$

for the conservation for a perfect fluid can be used to simplify the left-hand side of (4.21). And when that is done we get

$$-V_H d\rho = \frac{c^2}{G} dR_H,$$

which can be integrated to give the Friedmann-Lemaître equation,

$$H^2 = \left(\frac{\dot{a}}{a}\right)^2 = \left(\frac{8\pi G}{3}\right) \rho, \tag{4.23}$$

if the integration constant is set equal to zero, where a is the cosmic scale factor.

The density of a perfect gas remains constant, and so (4.22) reduces to $\rho = -p/c^2$, which is also the condition that the dark energy remain constant [DAESa]. Hence, (4.22) is *not* the first law if ρ is interpreted as the mass density. Alternatively, if ρ is interpreted as the internal energy density, then (4.22) is an expression of the first law, and using it to evaluate the time derivative of (4.23) results in the second field equation,

$$\frac{\ddot{a}}{a} = -\frac{4\pi G}{3}(\rho + p/c^2).$$

And neither is (4.21) the second law. The first law is $dE + pdV = 0$, and *not* (4.20). In fact, EFS use this to derive (4.22) with $E = Mc^2$. This is in blatant contradiction with their first law, (4.20). The second law, if it existed, would be $dE + pdV = TdS$, and *not* (4.21). Moreover, at zero chemical potential, the entropy density, $s = h = \varepsilon + p$, where h is the enthalpy density and ε is the internal energy density. However, for the putative expression for the entropy of a black hole, (4.14), an entropy density does not exist.

Bibliography

[DAESa] P. H. Frampton D. A. Easson and G.F. Smoot. Entropic accelerating universe. Technical Report. arXix:hep-th/1002.4278v3.

[DAESb] P. H. Frampton D. A. Easson and G.F. Smoot. Entropic inflation. Technical Report. arXix:hep-th/1003.1528v3.

[Ein11] A. Einstein. On the influence of gravitation on the propagation of light. *Ann. der Phys.*, 35, 1911.

[E.V] E. Verlinde. On the origin of gravity and the laws of motion. Technical Report. arXiv:hep-th/1001.0785.

[Lav95] B. H. Lavenda. *Thermodynamics of Extremes*. Horwood, Chicester, 1995.

[LL58] L. D. Landau and E. M. Lifshitz. *Statistical Physics*. Pergamon Press, Oxford, 1958.

Chapter 5

What is the Vacuum?

5.1 How to Unify General Relativity, Quantum Mechanics and Thermodynamics

Hawking is accredited with the unification of general relativity, quantum mechanics, and thermodynamics by attributing a temperature, T_H, to a black hole by setting it proportional to the surface gravity, κ,

$$T_H = \frac{\hbar\kappa}{2\pi ck},\tag{5.1}$$

of a Schwarzschild black hole. The identification was achieved through the Planck factor of black body radiation. But appearances can be deceiving.

In a series of papers [Unr76, Ful73, Dav75] (5.1) was generalized to the acceleration of any observer with respect to the 'zero-temperature Minkowski vacuum',

$$T_U = \frac{\hbar a}{2\pi kc},\tag{5.2}$$

in flat spacetime, where a is the acceleration in the instantaneous rest frame of the detector.

This was accomplished by showing that accelerations cause Doppler shifts, and the association of temperature T_U with acceleration a is again done by associating the acceleration in the Planck factor,

$$\frac{1}{e^{2\pi c\Omega/a} - 1},\tag{5.3}$$

with the temperature, where Ω is the frequency. Then, if the temperature, (5.2), measures thermal radiation, the acceleration must be given by $a = c^2/\lambda$, where λ is the wavelength. It is also crucial to observe that the density of states is missing from (5.3), and cannot be derived from thermal field theory. Equation (5.3) is, therefore, one of appearances rather than one of substance.

In order to get an idea of the magnitude of (5.2), consider the equation of state an ideal gas, $p = nkT$, where the pressure is due to gravity $p = \mu g$, with μ the mass per unit area [Jea19]. Evaluating (5.2) with this relation results in

$$\frac{\mu}{n} = \frac{\hbar}{2\pi c}.$$

The left-hand side is the mass times the wavelength, while the right-hand side identifies the wavelength as the Compton wavelength. Setting this wavelength equal to the ultra-relativistic thermal wavelength gives temperatures of the order of magnitude of rest energy of particle in question. And further setting the gravitational mass equal to $\hbar\omega/c^2$ leads to a linear relation between temperature and frequency so that (5.2) will have the formal appearance of Wien's displacement law.

Shortly after Planck's discovery of black body radiation, the effect of uniform motion on cavity radiation was investigated by his student Mosengeil in order to determine the dependencies of thermodynamic quantities on the motion, and published by Planck after his premature death in a mountaineering accident.

Planck found that both the quantity of heat absorbed and the temperature decreased when in motion so as to leave their ratio — the entropy — constant, and that the pressure was also an invariant. That the entropy be independent of the uniform motion was necessary for, otherwise, the degree of disorder of a system could be used to distinguish between a state at rest and one in uniform motion.

However, the Unruh temperature, (5.2), predicts that through the uniform acceleration of the cavity, a constant temperature of the cavity walls can be created and maintained. According to Unruh, such cavity radiation "is real enough to roast a steak" [MOSC04]. It will become apparent that a hungry man will literally starve to death before his steak gets cooked!

In this chapter we will show that both the Hawking-Unruh temperatures, (5.1) and (5.2), respectively, as well as the hypothesized thermal radiation that they are supposed to measure, are complete nonsense.

5.2 Can Acceleration Cause Doppler Shifts?

Almost a decade after the advent of the relativity of uniform motion by Poincaré, and subsequently by Einstein, Born [Bor09] generalized it to uniform acceleration. The motion was referred to as hyperbolic motion because it involved hyperbolic functions.

Born's point of departure was the expression of the ratio of the relativistic force, F, to the rest mass m_0:

$$\frac{d}{dt}\left(\frac{v}{\sqrt{1 - v^2/c^2}}\right) = \frac{F}{m_0} = a, \quad v = \frac{dz}{dt}. \tag{5.4}$$

For the initial condition $v = 0$ at $t = 0$, a single integration of (5.4) gives

$$\frac{v}{\sqrt{1 - v^2/c^2}} = at.$$

Since the speed of propagation of light is finite, it is necessary to take into account two times: a lab, or coordinate, time t, and a 'local', in the terminology of Lorentz, or proper, time τ. The latter is the time that one would register if he or she could ride piggy-back on the body in relative motion. Their infinitesimal increments are related by

$$d\tau = \sqrt{1 - v^2/c^2}\, dt. \tag{5.5}$$

However, since the velocity is no longer constant, (5.5) cannot be integrated to give

$$\tau = \sqrt{1 - \frac{v^2}{c^2}}\, t, \tag{5.6}$$

which represents time dilation, or the slowing down of clocks in relative motion.

The velocity, v, can be expressed in terms of proper time as

$$v = c \tanh \frac{a\tau}{c}, \tag{5.7}$$

which upon inversion gives

$$a\tau = c \tanh^{-1}(v(\tau)/c) = \frac{c}{2}\ln\left(\frac{c + v(\tau)}{c - v(\tau)}\right). \tag{5.8}$$

By exponentiating both sides of (5.8),

$$e^{a\tau/c} = \sqrt{\frac{c + v}{c - v}}, \tag{5.9}$$

it would appear that uniform acceleration causes a longitudinal Doppler shift. And the Doppler shift increases (exponentially) with the proper time!

Since the right-hand side of (5.9) is equal to the ratio of the observed frequency to the frequency of emission, the left-hand side would imply that it is time dependent. This not unknown: in optics, this is known as chirping or a sweep signal, as it is sometimes referred to. Surface acoustic waves are frequently used to generate chirped signals, which can also be exhibited in ultrashort laser pulses. However, they cannot be obtained from a simple Doppler shift, as (5.9) would imply.

If the left-hand side of (5.8) were the hyperbolic measure of the velocity,

$$\bar{v} = c \tanh^{-1}(v/c),   \tag{5.10}$$

its linearity in the same z-direction would lead to

$$\bar{v}_3 = \bar{v}_1 + \bar{v}_2 = \tanh^{-1}\frac{v_1}{c} + \tanh^{-1}\frac{v_2}{c} = \tanh^{-1}\left(\frac{v_1/c + v_2/c}{1 + v_1 v_2/c^2}\right),   \tag{5.11}$$

which is just the Poincaré addition law for velocities,

$$v_3 = \frac{v_1 + v_2}{1 + v_1 v_2/c^2}.   \tag{5.12}$$

It was Einstein who generalized (5.12) to non planar velocities.

The same additivity for the hyperbolic velocity, (5.11), must also apply to the velocity, $\bar{v} = a\tau$. For an inertial system it is simply:

$$\tau_3 = \tau_1 + \tau_2 = \sqrt{1 - v^2/c^2}\,(t_1 + t_2) = \sqrt{1 - v^2/c^2}\,t_3.   \tag{5.13}$$

However, since the velocity is *not* constant, (5.5) cannot be integrated to give a second order Doppler effect, (5.6), but, rather

$$a\tau = c \sinh^{-1}\frac{at}{c}.   \tag{5.14}$$

Acceleration has destroyed a strict proportionality between proper time, τ, and coordinate time, t, and so has destroyed the second order Doppler effect.

The velocity addition law now leads to the conclusion that:

$$a\tau_3 = a\tau_1 + a\tau_2 = c\left\{\sinh^{-1}\frac{at_1}{c} + \sinh^{-1}\frac{at_2}{c}\right\}   \tag{5.15}$$

$$= c\ln\left[\left(\frac{at_1}{c} + \sqrt{\left(\frac{at_1}{c}\right)^2 + 1}\right)\left(\frac{at_2}{c} + \sqrt{\left(\frac{at_2}{c}\right)^2 + 1}\right)\right] \neq c\sinh^{-1}\frac{at_3}{c}.$$

$$\tag{5.16}$$

Since there is no second order Doppler effect, or time dilation, the proper and coordinate times will not be additive, as in (5.13).

Integrating (5.7), by taking into account $dt/d\tau = \cosh(a\tau/c)$, gives

$$z(\tau) = \frac{c^2}{a} \cosh \frac{a\tau}{c}, \qquad (5.17)$$

which together with the inverse of (5.14),

$$t(\tau) = \frac{c}{a} \sinh \frac{a\tau}{c} \qquad (5.18)$$

are the so-called Rindler coordinates.

The ratio of (5.17) to (5.18) should be a good measure of the velocity, but instead, it is precisely the inverse of the velocity, (5.8); in other words

$$\frac{z(\tau)}{t(\tau)} = c \coth \frac{a\tau}{c} = \frac{c^2}{v(\tau)}. \qquad (5.19)$$

This is because the initial condition should have been $z(0) = 0$ at $\tau = t = 0$, and not $z(0) = c^2/a$, in which case (5.17) should have been:

$$z(\tau) = \frac{c^2}{a} \left(\cosh \frac{a\tau}{c} - 1 \right). \qquad (5.20)$$

The ratio of (5.20) and (5.18)

$$z(\tau)/t(\tau) = c \tanh (a\tau/2c),$$

can be considered as a velocity, but with an argument only *half* that would expected for uniform motion. In contrast to (5.17), (5.20) does give the correct non-relativistic limit, $z \approx \frac{1}{2}at^2$ for $a\tau \ll c$, where the distinction between proper and coordinate times disappears because the speed of light is considered as infinite.

Even if time dilation, (5.6), were valid, expression (5.9) would still make no sense. Time dilation, (5.6), can be thought of as an expression of the arithmetic-geometric mean inequality, as we will now show.

A light signal is sent out at time t_A, reflected at some later time, t_R, and returns to where it was sent out initially in time t_B. The coordinate, or (arithmetic) average, time, and distance traveled are

$$t = t^a = \frac{1}{2} (t_A + t_B), \quad \text{and} \quad z = \frac{1}{2}c (t_B - t_A), \qquad (5.21)$$

respectfully, with

$$v = \frac{z}{t}. \qquad (5.22)$$

Adding and subtracting (5.21), in turn, give

$$t_B = (1 + v/c)t, \quad \text{and} \quad t_A = (1 - v/c)t.$$

Their product gives time dilation when the proper time is identified as the geometric mean time $\tau = t^g \equiv \sqrt{t_A t_B}$, while the square root of their ratio gives the longitudinal Doppler shift.

In contrast, if (5.8) were valid, it would imply

$$a\sqrt{t_A t_B} = \frac{1}{2} c \ln \frac{t_B}{t_A}, \quad \text{and} \quad \sqrt{1 - v^2/c^2}\, at = \ln \left(\frac{1 + v/c}{1 - v/c} \right). \qquad (5.23)$$

These expressions are clearly wrong; for example, consider the second equation and let $v \to c$ for which the left-hand member would vanish, while the right-hand member would tend to infinity.

Moreover, if we define hyperbolic time as

$$\bar{t} = t_0 \ln \frac{t}{t_0}, \qquad (5.24)$$

where $t = t_B$ is any generic time, and $t_0 = t_A$ is an absolute constant whose numerical value will depend on the arbitrary choice of the unit of time, an exponential Doppler shift [Lav11a],

$$e^{\bar{v}/c} = \sqrt{\frac{c + v}{c - v}}, \qquad (5.25)$$

is obtained in terms of the hyperbolic measure of the velocity, $\bar{v}$.

Taking the logarithm of both sides of (5.25) and introducing (5.24) give Hubble's law [Lav11a, p. 422],

$$\bar{v} = \frac{c\bar{t}}{t_0} = H\bar{z}, \qquad (5.26)$$

where $\bar{z} = c\bar{t}$, and $t_0 = 1/H$ is the Hubble period. In contrast, (5.8) gives $H\bar{z} = a\tau$, but Hubble's law is known not to depend on acceleration.

A comparison of (5.9) and (5.25) clearly shows that accelerations and velocities have been confused. Whereas the former belongs to Euclidean geometry, the latter are part of hyperbolic geometry, a type of non-Euclidean geometry in which the curvature is constant and negative.

5.3   Are There Time-Dependent Doppler Shifts?

A time-dependent phase can also be obtained directly by considering a standard non accelerated Minkowski plane wave,

$$\varphi_{\pm}(\tau) = Kz(\tau) \mp \omega_K t(\tau) = \pm \omega_K t(\tau) \left(1 \pm \coth \frac{a\tau}{c} \right) = \frac{\omega_K c}{a} e^{\pm a\tau/c},$$

$$(5.27)$$

where $K = \omega_K/c$, and the Rindler coordinates, (5.17) and (5.18), have been introduced.

Rather than exponentiating both sides of (5.8), we can write it as the Lorentz transform,

$$\omega'_K = \frac{\omega_K \mp Kv(\tau)}{\sqrt{1 - v^2(\tau)/c^2}} = \frac{\omega_K \left[1 \mp \tanh \frac{a\tau}{c}\right]}{\sqrt{1 - \tanh^2 \frac{a\tau}{c}}} = \omega_K e^{\mp a\tau/c}, \qquad (5.28)$$

for a frequency ω_K, and a wave vector $\mathbf{K}$, "parallel or anti-parallel to the z direction along which the observer is accelerated" [AM04], i.e., $K = \pm \omega_K/c$. Equation (5.28) should correspond to the phase of the plane wave (5.27) when it is multiplied by the coordinate time, $t(\tau)$. That it doesn't means that neither the Lorentz transform, nor plane wave solutions that lead to a time-dependent phase, can be applied to non-inertial systems. This was not unexpected.

Furthermore, for $K = 0$, (5.28) reduces to a second order Doppler shift,

$$\omega'_0 = \omega_0/\sqrt{1 - v^2/c^2}, \qquad (5.29)$$

which we know does not exist because the proper and coordinate times are related by (5.14) and not by (5.6).

For small values of $a\tau$, (5.28) reduces formally to the familar Doppler shift, $\omega'_K \simeq \omega_K (1 \mp a\tau/c)$. However, it now involves *time-dependent* Doppler shifts detected by an accelerated observer. Thus, the fractional Doppler shift,

$$\frac{\Delta\omega}{\omega} \equiv \frac{\omega'_K - \omega_K}{\omega_K} = \mp \frac{2\pi kT}{\hbar}\tau, \qquad (5.30)$$

should be proportional to the temperature on the strength of (5.2).

The classical Doppler shift for an emitter approaching an observer with velocity v is

$$\frac{\Delta\omega}{\omega} = \frac{v}{c}. \qquad (5.31)$$

If the atoms are in equilibrium at a temperature T, their velocity distribution will be Maxwellian, and (5.31) can be used to determine the corresponding frequency distribution.

If ω_0 and $\omega_{\frac{1}{2}}$ denote the value of the frequency for which the intensity is maximum and at half-maximum, respectively, then[1]

$$\frac{\Delta\omega}{\omega_0} \equiv \frac{\omega_0 - \omega_{\frac{1}{2}}}{\omega_0} = \sqrt{\frac{2}{3}\ln 2}\,\sqrt{\frac{\langle v^2\rangle}{c^2}} = \sqrt{\frac{2}{3}\ln 2}\,\sqrt{\frac{3kT}{mc^2}}, \tag{5.32}$$

where $\sqrt{\langle v^2\rangle}$ is the root-mean-square velocity. Consequently, the classical Doppler shift is proportional to the *square root* of the temperature, and not to the temperature itself, as (5.30) would imply, with a proper time left over.

In contrast, a second order Doppler shift is proportional to the temperature for speeds $\sqrt{\langle v^2\rangle} \ll c$ [c.f., (5.66) below]. And, as we have previously mentioned, there is neither rhyme nor reason why a classical Doppler shift should increase with proper time.

Just as no temperature can be defined from translational motion alone because all inertial frames are equivalent, the principle of equivalence would also prohibit a temperature resulting from uniform acceleration alone.

Now consider a plane wave to be propagating in the $-z$ direction, with a time-dependent phase given by (5.27),

$$\phi_+(\tau) = \frac{\omega_K c}{a}e^{a\tau/c}.$$

The spectral distribution will be the modulus square of the Fourier transform of $e^{i\phi_+(\tau)}$:

$$|A|^2 = \left|\int_{-\infty}^{+\infty} e^{i\Omega\tau + i\omega_K c e^{a\tau/c}/a}\,d\tau\right|^2. \tag{5.33}$$

The double exponential integral,

$$A = \int_{-\infty}^{+\infty} e^{i\Omega\tau}e^{i(\omega_K c/a)e^{a\tau/c}}\,d\tau = \int_0^\infty x^{i\Omega c/a-1}e^{i(\omega_K c/a)x}\,dx, \tag{5.34}$$

can be reduced to a Gamma integral,

$$A = \frac{c}{a}\left(\frac{a}{\omega_K c}\right)^{i\Omega c/a}\Gamma\left(\frac{i\Omega c}{a}\right)e^{-\pi\Omega c/2a}, \tag{5.35}$$

through the substitution $x = e^{a\tau/c}$.

---

[1] The line breadth can also be expressed in terms of the 'most probable velocity', $\sqrt{2kT/m}$.

Now employing the formula,

$$\left| \Gamma\left(\frac{i\Omega c}{a}\right) \right|^2 = \frac{\pi}{(\Omega c/a)\sinh(\pi\Omega c/a)},$$

the spectral density is found to be

$$|A|^2 = \frac{2\pi c}{\Omega a} \frac{1}{e^{2\pi\Omega c/a} - 1}. \tag{5.36}$$

This leads to the unavoidable conclusion that the time-dependent Doppler shift detected by a uniformly accelerating observer has a thermal Planck spectrum with a temperature given by (5.2). But, it's better not to jump to any hasty conclusions, even if they appear to be unavoidable.

The Planck factor, (5.3), is only half the story. Missing is the correct density of states so that when (5.36) is integrated over all frequencies, it will give Stefan's T^4-law, which we know it must. The density of states given in (5.36) is wrong. This follows from the integral,

$$\int_0^\infty \frac{x^{\nu-1}dx}{e^{\mu x} - 1} = \frac{1}{\mu^\nu}\Gamma(\nu)\zeta(\nu), \tag{5.37}$$

where ζ is the Riemann zeta function, which has the domain of definition $\Re\,\mu > 0$ and $\Re\,\nu > 1$ [GR80]. Since the spectral density (5.36) has $\nu = 0$, the integral over all frequencies will not converge so that Stefan's law will not be obtained. The correct density of states is $8\pi\Omega^2/c^3$, and multiplying it by the energy $\hbar\Omega$ and the Planck factor, (5.3), give Stefan's law when integrated over all positive frequencies.

5.4 What's Wrong with Thermal Field Theory?

Thermal field theory is constructed from (second) quantized oscillators, and the space is known as Fock space. Hawking derived his black body distribution for radiation emanating from a black hole by evaluating the coefficients of a thermal Bogoliubov transformation between two asymptotic configurations 'in' and 'out'. It is therefore not surprising that Hawking came up with a Planck factor, (5.3). But what is surprising — and wrong — is what he identified the temperature to be. We will do so by showing that the thermal field theory itself is flawed because it can never account for the correct density of states, no matter the dimensionality of the system.

In thermal field theory, one starts with a discrete set of wave vectors $\mathbf{k}$ and then passes to a continuum according to

$$\sum_{\mathbf{k}} \longrightarrow \int d^3 k.$$

This necessitates expressing quantities like the spectral density as

$$\rho(\omega) = \hbar\omega \times \bar{n}(\omega) \times \frac{\omega^2}{\pi^2 c^3}, \tag{5.38}$$

a product of the energy of an oscillator, $\hbar\omega$, the particle number,

$$\bar{n}(\omega) = \frac{1}{e^{\beta\hbar\omega} - 1}, \tag{5.39}$$

where β is inverse temperature, and the number of oscillators between ω and $\omega + d\omega$, $\omega^2/\pi^2 c^3$. The particle number (5.39) is what enabled Hawking to associate $8\pi M$ with β, where M is the mass of the black hole.

However (5.39) has no meaning. It is derived from the expression for the probability of their being n particles in a single state,

$$P_n = \frac{\bar{n}^n}{(\bar{n}+1)^{n+1}} = \frac{e^{-n\beta\hbar\omega}}{\sum_{n=0}^{\infty} e^{-n\beta\hbar\omega}}. \tag{5.40}$$

The probability distribution (5.40) can be written as an error law for which the average $\bar{n}$ is the most probable value of n that is measured [Lav91]. The error law,

$$P_n = \exp\{S(n) - S(\bar{n}) - S'(\bar{n})(n - \bar{n})\}, \tag{5.41}$$

shows that the thermodynamic, and random, entropies $S(\bar{n})$ and $S(n)$, respectively, are sufficient to determine the form of the error law, where the prime means differentiation with respect to the argument. The property of concavity,

$$S(n) - S(\bar{n}) - S'(\bar{n})(n - \bar{n}) \leq 0,$$

ensures that (5.41) is a proper probability distribution, i.e., $P_n \leq 1$.

There is a general theorem by Callen and Greene [CG52] to the effect that there should be a single thermodynamics for all ensembles. This necessitates that the random entropy, $S(n)$, be the same function of the fluctuating variable, n, that the thermodynamic entropy, $S(\bar{n})$ is of its average value, $\bar{n}$. The entropy,

$$S(\bar{n}) = (1 + \bar{n})\ln(1 + \bar{n}) - \bar{n}\ln\bar{n}, \tag{5.42}$$

would identify the Poisson distribution if Stirling's approximation could be used, but it can't because there is only one cell to occupy. This is borne out by introducing (5.42) into Gauss's principle, (5.41); it will only reproduce the probability distribution, (5.40), if the random entropy vanishes.

Since the number of cells has shrunk to one, a random entropy, $S(n)$, cannot be defined. This is also true of (5.42) since thermodynamics applies when $\bar{n} \gg 1$, and (5.42) is effectively zero. This also means that (5.40) destroys the equivalence of the microcanonical and canonical ensembles, and is another blemish on thermal field theory that attempts to construct a thermal vacuum from single harmonic oscillators in every frequency interval.

This is exemplified by the fact that the total entropy is not the triple integral,

$$S \neq \int d^3k\{(1 + \bar{n})\ln(1 + \bar{n}) - \bar{n}\ln\bar{n}\}, \tag{5.43}$$

contrary to what has been claimed [Ume93]. Rather, the total entropy is the integral over all frequencies of the entropy of the negative binomial distribution,

$$S(\bar{n}) = (m + \bar{n})\ln(m + \bar{n})\ln(m + \bar{n}) - \bar{n}\ln\bar{n}, \tag{5.44}$$

where $m = \omega^2/\pi^2c^3$ is the number of oscillators in the frequency interval $d\omega$. This number has to be large enough so that Stirling's approximation applies.

Provided Stirling's approximation is applicable, Gauss's principle (5.41) will give the probability distribution as

$$P_n = \frac{(m + n)^{m+n}}{n^n} \cdot \frac{\bar{n}^n}{(m + \bar{n})^{m+\bar{n}}} \simeq \binom{m + n - 1}{n} p^n q^m, \tag{5.45}$$

which clearly shows the role of Stirling's approximation. The first term in the last expression of (5.45) is the negative binomial coefficient for $n + m \gg 1$, and the probabilities for success and failure are $p = \bar{n}/(m+\bar{n}) = e^{-\beta\hbar\omega}$ and $q = m/(m + \bar{n}) = 1 - e^{-\beta\hbar\omega}$, respectively. The particle density, obtained from the second law, $\partial S/\partial\bar{n} = \hbar\omega/T$, at zero chemical potential, is

$$\bar{n} = \frac{m}{e^{\beta\hbar\omega} - 1}, \tag{5.46}$$

in any frequency interval, and not (5.3). The chemical potential vanishes because photons are not conserved.

The distinction between radiation (photons) and (Planck) oscillators can be made considering the zero-point energy, and whether it should be

integral or half integral, like in Planck's second theory. The need for postulating semi-integral values of the zero-point energy arose when the Rayleigh-
Jeans limit was imposed on Planck's spectral distribution so that the law of
equipartition should be validated. But, the addition of a term linear in the
frequency to Planck's law would result in an ultraviolet catastrophe anew.

Yet, it arises naturally in the second quantization of oscillators whose
Hamiltonian is

$$H = \frac{1}{2}(p^2 + \omega^2 q^2).$$

When the momentum, p, and displacement, q, are expressed in terms of cre-
ation, $a^\dagger$, and annihilation, a, operators, according to $q = i\sqrt{\hbar/2m\omega}(a - a^\dagger)$
and $p = \sqrt{m\hbar\omega/2}(a + a^\dagger)$, the Hamiltonian, H, acquires a half-integral zero
point energy as a result of the commutation relation, $[a, a^\dagger] = 1$. Nevertheless the spectral densities can only change by integral units, and since these
are related to probability distributions it implies that neither photons nor
the Planck oscillators can be split in two. Therefore, half-integral zero-point
energies do not correspond to physical reality.

In the Einstein-Hopf theory, dissipative effects were accounted for in the
over-damped case where the acceleration has died out so that the force F
becomes proportional to the velocity v, according to the phenomenological
relation $F = -Rv$. The resistance, R, is to be determined by the fluctuation-
dissipation relation [c.f., §1.2.1],

$$\Delta^2/\delta t = 2RkT,$$

where δt is the time interval.

Einstein and Hopf, and later Einstein and Stern, did not have any prescription for determining the dispersion in energy, Δ^2, except for the fact
that

$$R = \frac{4\pi^2 e^2}{15mc^2}\left[3\rho(\omega) - \omega\frac{d\rho}{d\omega}\right],$$

should result in Planck's law. In contrast, if we know the entropy then the
dispersion is merely

$$-\left(\frac{\partial^2 S}{\partial \bar{n}^2}\right)^{-1} = \Delta^2.$$

Einstein and Hopf wrote

$$3\rho(\omega) - \omega\frac{d\rho}{d\omega} = \frac{1}{kT}\rho \cdot U,$$

where U is the oscillator energy. When they solved the differential equation they found the Rayleigh-Jeans law, valid in the low frequency limit. By substituting $U + \hbar\omega$ for U they found Planck's spectral distribution subject to $\rho(0) = 0$. With the entropy of the negative binomial distribution, (5.44), the dispersion is

$$-\left(\frac{\partial^2 S}{\partial \bar{n}^2}\right)^{-1} = \bar{n} + \frac{\bar{n}^2}{m}.$$

Einstein was quick to interpret the first term as arising from the particle aspects of radiation while the second term represented its wave-like properties.

Rather, had they replaced U by $U - \hbar\omega$, they would have come out with an average particle density given by

$$\bar{n} = m\left(\frac{1}{e^{\beta\hbar\omega} - 1} + 1\right) = \frac{m}{1 - e^{-\beta\hbar\omega}}, \tag{5.47}$$

which differs from Planck's law by an integral zero-point. Its physical interpretation implies that there should be at least one particle in each of the cells whose distribution is governed by the Pascal distribution [Lav91],

$$P_n = \binom{n-1}{m-1} p^n q^{n-m}, \quad n = m, m+1, \ldots$$

There is yet another distinction with the negative binomial distribution (5.45) insofar as m must be an integer meaning that it can also be negative. Since $|m|$ is very large, the integer part is of little consequence, but since m varies as the cube of the frequency, (5.47) is the negative frequency counterpart of (5.46). This recalls Schrödinger's idea of associating particle production with negative frequencies.

5.5 Bag Model of the Vacuum

At zero temperature, and at nuclear densities, quarks and gluons are confined in hadrons. Free quarks, or free gluons, are not free to propagate in the vacuum.

Quantum chromodynamics (QCD) predicts that at a high enough temperature all this may change; a first order phase transition may occur leading to a gas of free quarks and gluons, forming a quark-gluon plasma. According to the big-bang, this state of matter supposedly subsisted for a time $10^{-6}-10^{-5}$s after the bang. Such a state of matter could also exist in the core of neutron stars, and in collisions of ultra relativistic particles.

Such a phase transition can be investigated in a rather crude model which attempts to incorporate two basic features of QCD: asymptotic freedom, and confinement. Hadrons are conceived as 'bubbles,' or 'bags,' in a vacuum [ACW74]. Inside the confining bag, quarks are allowed to move freely. This is achieved by applying a constant pressure, B, to the bag so as to prevent quarks and gluons from escaping.

The total energy is the sum of the ultra relativistic kinetic energy and the compressive work involved in the confinement,

$$E = \frac{C}{R} + \frac{4\pi}{3}B\,R^3, \tag{5.48}$$

where the uncertainty principle, $p\,R \sim \hbar$ has been used to obtain the first term, and C is a constant. To find the radius of the spherical bag, R, (5.48) is to be minimized with respect to the bag radius, R. The stationarity condition is used to fix the radius at

$$R = \left(\frac{C}{4\pi B}\right)^{1/4}. \tag{5.49}$$

The total pressure is defines as

$$P = -\frac{\partial E}{\partial V} = \frac{C}{4\pi R^4} - B. \tag{5.50}$$

At equilibrium, (5.50) vanishes: the contribution of the quark kinetic energy is balanced by the contribution $-B$, due to the vacuum, as (5.49) shows.

Eliminating B between (5.48) and (5.49) results in $E = \frac{4}{3}C/R$. The ultra relativistic kinetic energy can be thought of as thermal radiation, $E_r = \sigma\frac{4\pi}{3}R^3T^4$, where σ is the Stefan-Boltzmann constant. This identifies (5.48) as the enthalpy, and $E_r = 3BV$, as the thermal equation of state of an ideal ultra relativistic gas. The condition,

$$\frac{C}{R} = \frac{4\pi}{3}\sigma R^3T^4, \tag{5.51}$$

determines the adiabatic condition, $RT = $ const., since C is a constant. Using (5.49) to eliminate C in (5.51) results in the 'critical' temperature,

$$T = \left(\frac{3B}{\sigma}\right)^{1/4}, \tag{5.52}$$

for the onset of a first order phase transition between a hadronic and plasma phase. Even though the transition is first order, particle physicists still speak of a critical temperature that is characteristic of a second, or order-disorder, transition.

However, the bag model is a little too naïve to be of any predictive power. The reason is that a constant pressure, B, implies a constant volume through the stationarity condition, (5.49). The adiabatic condition, $RT = $ const., then implies that the temperature is constant so that adiabats and isotherms coincide. According to Maxwell [Max04], the thermal properties of a substance can only be defined completely when they are specified on both isotherms and adiabats. However the two can only coincide at absolute zero. At any finite temperature, the slope of an adiabat is always steeper than the slope of an isotherm in the PV-plane. It is the assumption of a constant pressure, B, that destroys the predictive power of the bag model, and renders it completely useless [Lav07].

5.6 Unruh's Temperature and Wien's Displacement Law

According to Unruh, a particle accelerating in a vacuum would experience black body radiation surrounding it at a temperature directly proportional to its acceleration. By Einstein's equivalence principle, a particle at rest in a gravitational field would experience the same thermal radiation. This would fix the temperature to be zero absolute.

In this section we will show that the Unruh-Hawking temperature is simply Wien's displacement law in disguise, with the added feature in Hawking's case that the frequency is inversely proportional to the central mass.

The ratio

$$\frac{a}{2\pi c} = \omega, \tag{5.53}$$

defines a constant angular speed, ω, and introducing $2\pi c = \omega\lambda$ gives $a = \omega^2\lambda$, the centrifugal acceleration. It then follows that

$$\frac{\omega^2\lambda}{\omega\lambda T} = \frac{\omega}{T} = \text{const.}, \tag{5.54}$$

which is Wien's displacement law, where ω is necessarily the frequency at which the spectral density peaks. Unruh's law, (5.2), makes (5.54) a constant for *any* angular speed, ω, whereas Wien's displacement law strictly holds for that frequency at which the spectrum peaks, ω_{max}.

Since the speed of light is involved, Unruh's law can hardly be considered as one of uniform acceleration; rather, it is one of constant angular speed. The angular speed varies with temperature such that for a given T, $\omega = \omega_{max}$ is where the 'density' of the spectral distribution function has its maximum. The constraint of a constant angular speed, (5.53), determines the wavelength λ, and, consequently, the uniform centrifugal acceleration has reduced Unruh's law (5.2) to what formally looks like Wien's displacement law, (5.54).

5.7 Does Unruh Radiation Contradict the Equivalence Principle?

The perennial question as to whether a charge in a uniform gravitational field would radiate has seen its comeback many times. Looking at gravity as a force from the special relativistic viewpoint, it would appear that the answer is 'yes', but considering gravity as geometry from a general relativistic perspective one would be inclined to say 'no', which is what Bondi and Gold claimed.

Their conclusion was arrived at by invoking Einstein's equivalence principle which they claimed states that "it is impossible to distinguish between the action on a particle of matter at constant acceleration or of static support in a gravitational field." For if the accelerating particle did radiate it must also radiate in a static gravitational field, whose radiation:

> would reach distant regions unaffected by the gravitating body, where Maxwell's equations apply without modification due to gravitational effects. According to these equations there can be no static radiation field; and as the whole system is static the electromagnetic field cannot depend on time.

Hence, there will be no radiation.

It would therefore appear that the equivalence principle does not apply to charged particles. Others would disagree. Milionni [Mil94] sums up the prevailing mood as:

> It is now generally accepted that a uniformly accelerated charge does radiate and the radiation does not contradict the fact that the reaction radiation force vanishes during uniform acceleration. It has also been shown that the fact of radiation does not contradict the Principle of Equivalence in the general theory of relativity: an observer falling with a charge in a uniform gravitational field will detect no radiation.

The backdrop for a uniformly accelerating charge is flat Minkowski space, while that of a static charge is the Schwarzschild field. Now, there are a lot of contradicting elements afoot. The first is the equivalence principle which, in a loose sense, says that physical phenomena in a uniformly accelerating system are equivalent to the same phenomena occurring in a uniform gravitational field. This is because there is no force in general relativity so that Newton's second law is absent. Since in special relativity gravity is still a force, there is no common ground where general and special relativity meet. So there is a dichotomy here, and this is what the equivalence principle supposedly connects.

To every radiating body we can associate a temperature, but that temperature need not be the temperature of black body radiation, which Einstein showed resulted from a dynamic equilibrium between absorption and emission. Spontaneous emission alone resulted in Wien's distribution, whereas when combined with stimulated emission gave Planck's law.

Now, as we have seen, Unruh claims that when an observer is moving at constant acceleration in the vacuum he perceives himself to be immersed in a thermal bath at temperature (5.2). However, if the observer were in an inertial frame he would not measure any temperature or be aware of any radiation.

Yet, according to the equivalence principle, an accelerated charge is equivalent to a static charge in a gravitational field. Thus, the uniform acceleration in (5.2) can be replaced by the gravitational acceleration,

$$a = \frac{GM}{r^2}, \tag{5.55}$$

so that (5.2) becomes

$$T = \frac{\hbar GM}{2\pi kcr^2}, \tag{5.56}$$

which would be alright if r were chosen as the radius of the earth so that $GM/r^2 = g$. However, Hawking chose to evaluate r at the Schwarzschild radius, $r = 2GM/c^2$, so that (5.56) becomes (5.1) with surface gravity, $\kappa = c^4/4GM$. It is this constraint on r that dashes any hope for thermodynamic connection between mass-energy and temperature.

According to Milionni [Mil94], an accelerated observer should detect thermal radiation coming from the thermal field of the vacuum whereas a stationary observer, or one in a uniform gravitational field, should not detect anything. Consequently, a free-falling observer should detect no

Hawking radiation. According to Hawking, however, a "free-falling observer would not notice a large amount of particle production near the horizon," but he would notice some radiation! Also observe that by some miracle, thermal radiation has now become particle production — particles other than photons.

The role of the equivalence principle is to replace the acceleration of the observer by free-fall, and not by replacing one temperature by another one. An observer stationary on the surface of the earth would be equivalent to one in free fall at 1 g in a space ship. Smolin [Smo00] tells us to look at (5.2) as a kind of addendum to Einstein's equivalence principle.

> According to Einstein, a constantly accelerating observer should be in a situation just like an observer sitting on the surface of a planet. Unruh told us that this is true only if the planet has been heated to a temperature that is proportional to the acceleration.

Temperature, however, does not need — nor require — the acceleration of an observer in order to be measured. Have we lost all our physical intuition, only to be guided by deceptive mathematical formulas?

According to Jacobson, the two temperatures, (5.1) and (5.2), are not equivalent, but are related through a 'lapse' function, $N = \sqrt{1 - 2GM/c^2 r}$, which supposedly tells how much proper time corresponds to coordinate time in the Schwarzschild metric. The relation he proposes is

$$T_U = (1 - N^2)^2 \, T_H,$$

which reduces to gravitational acceleration, (5.55), so that it is no surprise that it gives the Hawking temperature at the horizon, (5.1), for $N = 0$, while it vanishes as $r \to \infty$, when $N = 1$. But, how is it possible that what was uniform acceleration now has acquired a radial dependency? This destroys any hope of drawing an equivalence of the two temperatures in terms of Einstein's equivalence principle.

Why should gravitational acceleration be uniquely confined to the Schwarzschild radius? Is that where we would expect the same radiation as the black body radiation picked up by a uniformly accelerating detector? The Unruh temperature in cgs units is $4.1 \times 10^{-23} a$, so that it would take an acceleration of $a = 10^{20} g$ to reach a temperature of $1°$K. We could claim an equivalence if the two temperatures were equal. The same temperature would give a Schwarzschild radius of 10^{-2}, which would correspond to a planet the size of the earth. The acceleration predicted by the Unruh temperature is 10^{20} times too large!

The Unruh temperature, (5.2), would therefore provide a way of distinguishing between an observer in uniform acceleration at 1g and one stationary on the surface of the earth. The former would measure a temperature of the vacuum whereas the latter would not. But aren't the two supposedly equivalent under the equivalence principle? Moreover, Unruh would also take exception to static Hawking radiation in which uniform acceleration has been replaced by surface gravity — even if it is at the Schwarzschild radius.

So if we stick to the equivalence principle, the Unruh temperature measures thermal radiation in a flat spacetime at uniform acceleration so the Hawking temperature should predict the same radiation in the static Schwarzschild metric. While there is nothing pathological with black body radiation, there is something very wrong with the Hawking temperature, (5.1), for it associates thermal radiation with a body of negative heat capacity. This is not a paradox; it is simply plain wrong, as we shall elaborate upon in the following chapter.

In view of the Bondi-Gold conclusion, if there is no radiation of a charged particle in a stationary gravitational field, there would be none for one in a state of uniform acceleration. Thus, neither the Unruh, nor the Hawking temperature, would measure radiation — thermal, or otherwise.

Having come this far, the reader is undoubtedly confused as to what is going on. The equivalence principle is simply a red-herring. Observe that the Unruh temperature contains both Planck's constant $\hbar$ and the speed of light, c. This means that it is both quantum and ultrarelativistic. The 2π should not appear in expression (5.2). And when it is removed, and c^2/λ is substituted for the acceleration, we come out with Wien's displacement law, (5.54), which is numerically

$$\lambda = \frac{0.228}{T}.$$

It is this wavelength, in cm, at which the black body spectrum peaks for a given temperature. Moreover, it affirms that there is no detector in uniform accelerated motion, and there is absolutely nothing to apply the equivalence principle to. The same acceleration would give the expression for the frequency of a black hole as

$$\nu = \frac{c^3}{4GM},$$

in violation of the relation between frequency and gravitational 'mass,' $M = h\nu/c^2$.

In order to see what's really afoot in Unruh's interpretation of a temperature, consider his claim that a thermal spectrum of sound waves should be gotten from a sonic horizon/Mach shock waves in trans-sonic fluid flow. His expression for the temperature is

$$T = \frac{\hbar}{k} \cdot \frac{\partial v_s}{\partial r},\tag{5.57}$$

where v_s is the speed of sound,

$$v_s^2 = \frac{\partial p}{\partial \varrho},$$

with p the pressure, and ϱ the density. Since (5.57) contains $\hbar$, but does not contain c, we expect that it will be a non-relativistic, quantum relation.

The characteristic inverse thermal time is $\sqrt{\partial p / \partial \varrho} / \lambda_T$, where λ_T is the non-relativistic thermal wavelength,

$$\lambda_T = \frac{\hbar}{\sqrt{mkT}}.$$

Solving for the temperature, and equating it with (5.57) in the form

$$T = \frac{\hbar}{k\lambda_T} \sqrt{\frac{\partial p}{\partial \varrho}},$$

we get

$$\lambda_T = \frac{\hbar}{mv_s} = \frac{\hbar}{\sqrt{mkT}}.$$

This is none other than the equation of state,

$$\frac{\partial p}{\partial \varrho} = kT,\tag{5.58}$$

of an ideal gas!

Hence, there is no motion associated with the Unruh temperature (5.57), and all his expressions for the temperature as a function of an acceleration reduce to expressions for the thermal wavelength. In the ultrarelativistic limit we get Wien's displacement law, (5.54), while in the non-relativistic limit it becomes the equation of state for an ideal gas, (5.58). There is nothing new under the sun.

If the uniformly accelerating detector were emitting radiation, there would be a rate of energy loss proportional to the square of its acceleration. However, for black body radiation the total intensity of emission varies as the temperature to the fourth power. According to Unruh's expression (5.2) this would lead to a power loss proportional to the acceleration to the fourth power — and not as Larmor tells us — the square of the acceleration. Furthermore, if the two power losses were to be equated, it would fix the acceleration at unimaginable magnitude of 10^{31} times greater than the surface gravity of the earth! These magnitudes are as unbelievable as the concepts themselves.

Hawking claims that black holes do radiate and at a temperature given by (5.1). However, more recently [c.f., Preface], he now claims that black holes do not exist, and the event horizon has to be replaced by an 'apparent' horizon, which temporarily imprisons matter and energy only to allow them to be set free at some later date, and in some chaotic fashion. Hawking has found the need for a change of heart because of the incompatibility of a 'classical' black hole and quantum mechanics, that was spurned by the 'information paradox' fiasco.

However, there are very precise and stringent conditions that allow quantum mechanics to manifest itself on a classical scale, and black holes do not meet the requirements. Quantum mechanics usually manifests itself on a classical scale at very low temperatures where gases become degenerate, i.e., they do not conserve particle number, and can, therefore, be thought of excitations rather than *bona fide* particles.

The change of heart arose from the nonsensical question of whether black holes conserve information or not. Information conservation is not the same as energy conservation, but, if it has something to do with (negative) entropy, the question of its conservation should never arise in any irreversible process. Anyone being swallowed up by a black hole would certainly have more to worry about than how much information is being lost. Moreover, the conservation of information, if it has any meaning at all, is not the same as the conservation of probability so that no 'firewalls' are necessary. A lot of time and effort have gone into useless polemics [Sus08].

Expression (5.1) predicts that the smaller the black hole the hotter it will be. Primordial black holes which are of the size of hadrons were formed at the time of the big bang, and have been emitting radiation on a time scale comparable to the life of the universe with a peak in the black body radiation at 14 MeV. Satellite borne detectors measured a gamma ray flux that fell off with energy as a power law, $E^{-2.5}$, without any evidence for a photon excess in the range of 14 MeV. Rather than being used as evidence against the radiation of primordial black holes at extremely high energies, as (5.1) predicts, it was turned around into the Hawking-Page upper bound on the number of primordial black holes whose radiation can be detected! This is more often the rule than the exception: a null result being converted into an upper limit on observability.

For whatever wavelength we choose for the expression of the electro-magnetic acceleration, c^2/λ, it must satisfy the condition that its product with the free-fall frequency, $\nu_f = \sqrt{G\varrho}$, must be less than the speed of light,

$$\lambda \nu_f < c. \tag{5.59}$$

Since $\varrho \sim M/\lambda^3$, inequality (5.59) simply states that the wavelength has to be greater than the Schwarzschild radius. For a neutron star this would imply a frequency cut-off of about 10^5 Hz. Rather, if we identify the wave-length in (5.59) with the thermal wavelength, this imposes that tempera-tures must be less than $\hbar c^3/kGM$, or that the Hawking temperature (5.1) is unattainable.

Field theories thrive on analogies, and therein lead to their ultimate discredit. What looks like something else doesn't mean that it is. By merely doubling the number of degrees of freedom, field theory hopes to account for the bulk properties of matter like thermodynamics does. As it applies to black holes one set of dynamical variables would apply to our world which would be this side of the event horizon, while another dual set would apply to the other side. This would allow one to speculate that the origins of the thermal degrees in our world are due to the presence of other universes which are totally disjoint with our world except for sharing a common vacuum. We are well within the realm of science fiction.

The woes of cosmology are primarily due to 'thermal field' theory. Hawk-ing made use of the Bogoliubov mixing coefficients to show that black holes are thermal objects. His result is based on the event horizon dividing the solution of the massless wave equation into two classes: those which propa-gate on the remote past null light cone to the event horizon and out to the future null light cone, and those which are trapped by the horizon and never make it out. This results in particle production, and the association of the ratio of the mixed Bogoliubov coefficients with a Boltzmann factor leads to a Planckian spectral density. The conclusion was unavoidable if we believe in thermal field theory: a black hole radiates with a thermal spectrum.

Davies soon afterwards repeated Hawking's calculation for the Unruh temperature of constant acceleration using a 'wedge' of Rindler space where a 'wall' replaces the 'event horizon', and the geodesics by curves of constant acceleration, which are however, not geodesics at all. It is now the reflecting wall that appears to an accelerated observer to radiate at a constant temperature given by (5.2). The common theme is that the effect of an accelerated detector is to promote zero-point fluctuations to thermal fluctuations. Such a promotion is, however, only wishful thinking.

Using field theory, any incoming wave packet, expressed in terms of creation and annihilation operators bounces off a horizon, or wall, and is converted into an outgoing wave packet. The ratio of the Bogoliubov mixing coefficients is an exponential whose exponent can be considered as a ratio of frequency to temperature. There is nothing particular to black holes except that the surface gravity appears in the exponent. The resulting distribution for Bose particles, i.e., photons, never gives the correct density of states so that it is a matter of appearances than one of substance.

If, as has been claimed, surface gravity is equivalent — in the sense of Einstein's equivalence principle — to uniform rectilinear acceleration in the Unruh temperature (5.2), then, based on Einstein's equivalence principle, it should hold for any type of uniform acceleration like that of uniform circular motion. However, the same methods that were used to associate uniform rectilinear acceleration with a Planck spectrum found that a uniformly rotating detector does not yield a Planck spectrum, but rather, one that depends on the tangential velocity of the detector.[2]

Although the association of acceleration and temperature is wrong, it does highlight that all forms of acceleration are not equivalent, thus sounding the death kneel to Einstein's equivalence principle. This can also be gleaned from the metrics themselves: a flat metric takes into account angular momentum, gravity needs a non-flat one. Not only does uniform circular motion not produce what formally looks like a Planck spectrum, it also dampened hopes of introducing an Unruh temperature for synchrotron fluctuations.

McDonald regards "the concerns of Bondi and Gold ... on radiation and the equivalence principle as precursors to the concept of Hawking radiation." But even before Hawking came to his conclusion that the radiation must be thermal, Ginzburg asserted that "neither a homogeneous gravitational field nor a uniformly accelerated reference frame can actually 'generate' free particles, especially photons." The problem lies not with the radiation — granted such radiation exists — but with the requirement that

---

[2]An analogous situation arises between uniform motion — and not uniform acceleration — in the linear Doppler shift and uniform rotation in the angular Doppler shift. Whereas the linear Doppler shift in the frequency is proportional to the product of the emitted frequency and the relative velocity between emitter and detector, the angular Doppler shift is proportional to the constant angular velocity of the rotating body from which circularly polarized light has been scattered, emitted or absorbed.

it be thermal radiation. The latter is imposed by thermal field theory which is far from being suited to describe a black hole.

5.8 Kepler's Third Law, Escape Velocities, and Luminosity of Black Holes

Centrifugal acceleration is determined by Kepler's third law,

$$g = \frac{GM}{R^2} = \omega^2 R, \tag{5.60}$$

which would not result in uniform acceleration were it not for the fact that R is evaluated at the event horizon of the black hole, $R = \mathcal{R} := 2GM/c^2$, the Schwarzschild radius. Thus,

$$\frac{\hbar g}{2\pi c} = \frac{\hbar \omega^2 \mathcal{R}}{2\pi c} = \frac{\hbar c^3}{8\pi GM} = kT. \tag{5.61}$$

However, *it is incompatible to use the escape velocity,*

$$c^2 = \frac{2GM}{R}, \tag{5.62}$$

to evaluate the radius R in Kepler's law (5.60) which results from a balance between the gravitational potential and twice the rotational kinetic energy, $(\omega R)^2$, per unit mass. It is tantamount to doubling g in (5.60) thereby transforming Kepler's law into an expression for the escape velocity. The angular speed is fixed at

$$\omega = \frac{c^3}{\sqrt{8GM}}, \tag{5.63}$$

so that the tangential velocity is $\omega \mathcal{R} = c/\sqrt{2}$.

Solving (5.61) for the inverse temperature,

$$\frac{dS}{dM} = \frac{1}{T} = k\frac{8\pi GM}{\hbar c^3},$$

and integrating gives the exact expression for the Bekenstein-Hawking entropy,

$$S = k\frac{4\pi GM^2}{\hbar c^3}, \tag{5.64}$$

where the arbitrary constant of integration has been set equal to zero. In natural units, (5.64) would be $\frac{1}{4}A$, where $A = 16\pi M^2$ is the surface area of the event horizon. The fact that the frequency should decrease with the increasing mass in accordance to (5.63), rather than increasing with it, makes (5.64) less than an entropy [Lav13].

Luminosity should increase with the mass of a star. Luminosity is defined as the product of energy flux and the surface area:

$$L = \sigma T^4 4\pi R^2, \tag{5.65}$$

where σ is the Stefan-Boltzmann constant. As we have seen, the surface area increases as the square of the mass, which if it were not for the Hawking temperature would be physically reasonable. But, because the Hawking temperature decreases with the mass, the luminosity of a black hole would, in fact, decrease as the square of the mass, $L \sim M^{-2}$. Thus, even though the black hole is radiating as a black body, its luminosity is decreasing as its mass increases!

5.9 Centrifugal Acceleration and the Pound-Rebka Experiment

Consider the Pound-Rebka experiment where the relativistic time dilation of the nuclei in a crystal lattice is due to thermal vibrations. The ratio of the shift in frequency when a nucleus emits a gamma ray in motion, ω, to when it is at rest, ω_0, is [She60]

$$\frac{\omega}{\omega_0} = \sqrt{1 - \langle v^2 \rangle / c^2} - 1 \approx -\frac{1}{2} \frac{\langle v^2 \rangle}{c^2} = -\frac{3}{2} \frac{kT}{mc^2} = -2.4 \times 10^{-15} T, \tag{5.66}$$

where $\langle v^2 \rangle$ is the mean square velocity, caused by the random thermal vibration of the lattice, of the Fe^{57} nucleus whose mass is m. The angular speed can be estimated from $\hbar\omega = k\Theta$, where $\Theta = 467°$ K is the Debye temperature. From this it follows that $\omega/2\pi \simeq 10^{13}$ s^{-1}. If the oscillation is harmonic, the maximum acceleration of the nucleus is 10^{16}g. The mean atomic distance between the nuclei would be about 10 Å.

Now, if accelerations were responsible for the shift,

$$a = \frac{\langle v^2 \rangle}{r} = \frac{4.3 \times 10^6}{r} T. \tag{5.67}$$

At room temperature, (5.67) would give an interatomic spacing of 10^{-2} Å, which is two orders of magnitude too small to be an interatomic distance. From this we conclude that velocity slows down clocks, not acceleration.

That higher temperatures lead to higher frequencies means that temperatures, as they influence the velocities of the radiating nuclei, cause clocks
to go slower, and the period of the Fe^{57} 'clock' to become longer.

The conclusion arrived at by the temperature-dependent experiments
is that accelerations of the order of 10^{16} g, due to lattice vibrations, produce no intrinsic frequency shifts in Fe^{57} nuclei to an accuracy exceeding
1 in 10^{13} [She60]. Consequently, there is no substance to either (5.67) or
(5.2), which would require accelerations of order 7.4×10^{22} g, even at room
temperature!

Gravity, unlike linear acceleration, also slows down clocks due to the fact
that the gravitational field of a body is inhomogeneous. The gravitational
field increases towards the surface causing time to run slower; it has nothing
to do with motion. As we have seen, uniform gravitational acceleration
destroys the second order Doppler effect because proper and coordinate
times are related by (5.14), and not according to (5.6), or equivalently, by
their frequencies (5.29).

5.10   Are Minkowski and Rindler Spacetimes Equivalent?

It is commonly acknowledged that Rindler spacetime is a wedge portion
of two dimensional Minkowski spacetime [Wal84]. Whereas both have flat
spacetimes (i.e., zero curvature), a Minkowski observer moves with constant
velocity while a Rindler observer travels at constant acceleration. Although
it is apparent that uniform acceleration is *not* equivalent to uniform motion,
it is not apparent that the Rindler metric is related to the Minkowski metric
by (5.77) below, so that it is merely the Minkowski metric itself!

The transformation from Minkowski coordinates (t, z) (two dimensional
case) to Rindler coordinates (v, u), in units where $c = 1$, is [Par77]:

$$t + z = \quad ue^v, \tag{5.68}$$

$$t - z = -ue^{-v}. \tag{5.69}$$

Adding and subtracting (5.68) and (5.69) give

$$t = u \sinh v, \quad \text{and} \quad z = u \cosh v. \tag{5.70}$$

A comparison of (5.70) with (5.18) and (5.17) shows that $v = a\tau$ and
$u = 1/a$. The latter is a pure constant. This is also evident from

$$\frac{dz}{dt} = \tanh v, \quad \text{and} \quad v = \tanh^{-1}(t/z), \tag{5.71}$$

so that

$$\frac{dz}{dt} = \frac{t}{z}. \tag{5.72}$$

Integrating (5.72) gives

$$z^2 - c^2 t^2 = u^2 > 0, \tag{5.73}$$

where the integration constant is determined from (5.68) and (5.69). The square of the 'length' of the two vector is both constant and negative. However, if we evaluate the left-hand size of (5.28) using (5.18) and (5.20) we find

$$z^2 - c^2 t^2 = -4\frac{c^4}{a^2} \sinh^2(a\tau/c), \tag{5.74}$$

which is certainly not constant, and of opposite sign to (5.73).

And since $u = $ const., there is no metric of the form [Par77]:

$$ds^2 = dt^2 - dz^2 = u^2 dv^2 - du^2, \tag{5.75}$$

but just

$$\frac{dv}{ds} = a, \tag{5.76}$$

which reduces to $ds = d\tau$ when the definition of v is introduced. The 'time coordinate', v is proportional to the proper time s, but it is also the velocity, $a\tau$. The metric (5.75) therefore reduces to

$$d\tau^2 = dt^2 - dz^2, \tag{5.77}$$

in which all vestiges of acceleration have disappeared. A coordinate transformation cannot be responsible for the transformation of a state of uniform motion into one of uniform acceleration!

We therefore conclude that *there is no flat spacetime metric which can account for uniform acceleration. The lack of conservation of momentum, or angular momentum, shows up in the transformation from the dot product to the more general inner product to which it is conformal to, and which necessarily requires a non-flat metric.*

The logarithm of the ratio of (5.68) and (5.69),

$$v = \frac{1}{2} \ln\left(\frac{1 + z/t}{z/t - 1}\right), \tag{5.78}$$

which is the second equation in (5.71). It shows that neither Doppler's principle nor causality is obeyed. This is because the second transformation in (5.70) is incorrect. As we have already mentioned [cf., (5.20)], it should be

$$z = u(\cosh v - 1),   \tag{5.79}$$

which is necessary in order that $z \approx \frac{1}{2}uv^2$ in the non relativistic limit. Consequently, the ratio

$$\frac{z}{t} = \tanh(v/2),   \tag{5.80}$$

so that

$$v = \ln\left(\frac{1 + z/t}{1 - z/t}\right) = \frac{1}{2}\ln\left(\frac{1 + dz/dt}{1 - dz/dt}\right).   \tag{5.81}$$

Hence, it is not true that [Rin66] "a Lorentz transformation applied to the Minkowski coordinates

$$z' = z\cosh\psi - t\sinh\psi, \qquad\qquad t' = -z\sinh\psi + t\cosh\psi,   \tag{5.82}$$

where

$$e^{2\psi} = \frac{1 + dz/dt}{1 - dz/dt},$$

induces the transformation"

$$u' = u, \qquad\qquad v' = v - \psi,$$

since $\psi = v$. And if $v \neq$ const., there will be no Lorentz transformation, (5.82), and no Doppler shift.

The transformation from Minkowski to Rindler space was supposedly necessary to get frequency mixing [Ful73]. Using Ehrenfest's adiabatic principle, Schrödinger [Sch39] showed that there would be no frequency mixing in a universe undergoing uniform expansion. According to him, this meant no particle production. But, Schrödinger alluded to the possibility of getting mixing if there was accelerated expansion. However, this would violate Ehrenfest's theorem, and the radiation would no longer be thermal.

On a number of occasions we have appealed to a theorem of Lord Rayleigh whereby a normal mode will retain its integrity if the dimensions of the enclosure are varied sufficiently slowly such that [Sch39]

$$\left|\frac{\dot{R}}{R}\right| \leq \omega,   \tag{5.83}$$

in order that the energies of all proper vibrations, ε_ω, change in the same measure as their frequencies, ω, i.e., $\varepsilon_\omega/\omega = $ const. On the contrary, in order for their to be pair production, the sign of the inequality (5.83) must be inverted [Par77]

$$\left|\frac{\dot{R}}{R}\right| \geq \omega. \tag{5.84}$$

In fact that same analogy that was used by Ehrenfest to derive his adiabatic theorem is used in requiring inequality (5.84) to hold. The frequency of the quantized field, ω, is the analog of the frequency of a pendulum. If a pendulum of length R is initially in its ground state, and its length is changed at a rate $\dot{R}$, then the probability of excitation will approach unity when the average frequency of the pendulum satisfies (5.84).

At the first Solvay conference, Einstein mentioned to Lorentz that the ratio of energy to the frequency would be unaffected if the length of the pendulum were slowly varied so that condition (5.83) would be satisfied [Kle70]. It is precisely this condition that we would expect the black body spectrum to remain unaltered so if Hawking-Unruh radiation exists at all, it cannot be thermal radiation.

5.11 Getting Mass from Nothing: The Higgs Mechanism

Gauge theories developed from two properties of the nuclear force: (i) the force between nucleons was known to have an extremely short range, as opposed to electromagnetism which has an infinite range, and (ii) the nuclear force was found to be charge-independent.

In regard to (i), Yukawa appended an exponential decay term onto the Coulomb potential which contained the mass of this new quantum. This introduced a cut-off for the interaction. Concerning (ii), Heisenberg invented isotopic spin indicating that since the nuclear force between neutron and proton was the same as that between two neutrons, or two protons, the proton and neutron could be considered as 'up' and 'down' states of the same particle, in analogy with the two spin states of the electron.

Almost a quarter of a century later, Yang and Mills postulated that the strong nuclear interaction is exactly gauge invariant like electromagnetism. The relation between the two is the same as the relation between the groups $U(1)$ and $SU(2)$, the unitary group and the special unitary group, respectively. The unitary group, $U(1)$, has one component — the phase — where the vector potential supplies the connection between the phases of the wave function at different positions. The special unitary group, $SU(2)$, is slightly more complicated having two members, not just one. It is the (covering) group of rotations in three dimensional space, and it can relate the 'up' component of a particle of isotopic spin $1/2$, at one point in space to an 'up' state at a different point in space.

What is essential to bear in mind is that the components of $SU(2)$, which can be expressed in terms of the Pauli spin matrices, are identical to the components of the Stokes vector which specifies the polarization of a particle [c.f., §1.1.6 and [Lav11b]]. However the components of the Stokes vector are analogous to the operators of angular momentum so the most general form of the Yang-Mills 'potential' is a linear combination of angular momentum operators.

The raising and lowering angular momentum operators, $L_+ = L_x + iL(y)$ and $L_- = L_x - iL_y$, respectively, are completely analogous to their Stokes counterparts, $U + iV$ and $U - iV$, while L_z is analogous to the difference between horizontal and vertical polarization, Q. The raising operator, L_+, for example, can transform a 'down' state into an 'up' state whereby a neutron absorbs a unit of isotopic spin from the gauge field and transforms itself into a proton. So in gauge jargon we may conclude that whereas the electromagnetic field does not transfer charge, the Yang-Mills gauge field does.

If we want to construct a Lagrangian from a classical Hamilton principle, the Lagrangian must not contain any mass, for a mass term of the form $m^2 A_\nu A^\nu$, where A_ν is the vector potential, would not leave the Euler-Lagrange equations invariant under a gauge transformation. The absence of this term in the Lagrangian means that the carrier of the field interaction must have zero mass, just as in the electromagnetic case where a photon has zero mass.

The question boils down to how do we get something from nothing? In this case that something is mass. An analogy had to be drawn, and it came from the field of solid state physics. The difference between a first and second order phase transition is that whereas the states are different at the transition point in a first order transition, they are identical in a second order one. This means that a second order phase transition is continuous in the sense that although the state of the body changes continuously, its symmetry, however, changes *discontinuously* at the critical point.

Moreover, the existence of two phases at the transition point means that there is a non-vanishing latent heat in a first order phase transition whereas it vanishes in a second order transition as a result of continuity. Thus, it is incorrect to invoke a second order phase transition in the case of inflation while claiming that the latent heat can reheat the universe to its original temperature prior to inflation. This shows that you can't have your cake and eat it too!

What changes discontinuously in a second order phase transition is the specific heat; it actually becomes infinite at the transition point. In the Landau-Ginsburg model, one defines an order parameter, ϕ, which is zero in the more symmetrical phase and different from zero in the less symmetrical phase. In the theory of the weak interaction, ϕ is referred to as a "new fundamental spin-0 field", aka the Higgs field.

The Landau-Ginsburg theory proceeds to develop the Gibbs free energy in terms of a Maclaurin series in ϕ beginning with the second order term because ϕ_0 is zero in the more symmetrical phase. This is because that, unlike the temperature and pressure which can be specified arbitrarily, the value of the order parameter, ϕ, must itself be determined from the condition of thermal equilibrium. Since ϕ_0 vanishes in the initial, more symmetrical phase the series expansion begins with a quadratic term, whose coefficient is fixed by a thermodynamic stability condition: it must be positive initially so that the more symmetrical state coincides with a minimum in the Gibbs free energy.

For symmetry reasons, the cubic term in the expansion must vanish leaving a quartic term: the probability for a displacement to the left must equal that to the right. The coefficient of the quadratic term changes sign as the transition point is crossed. So what was a minimum in the free energy at the origin has become a maximum afterwards. The condition for a finite value of the order parameter contains an arbitrary phase so that due to the invariance of the free energy, an arbitrary gauge transformation can rotate this ground state 2π, transforming the two valleys in the two dimensional plot of the free energy versus the order parameter into a Mexican sombrero in a three dimensional one.

Now particle physicists doctor up this picture somewhat. By associating the quadratic term in the series expansion of the free energy with the kinetic energy of the ϕ field, the condition for the symmetric minima in the free energy about the origin, which now corresponds to a maximum in the free energy, is related to a finite, positive value of the mass of the field. The 'true' vacuum is associated with the more symmetrical state in which the particles are massless, while the 'false' vacuum has particles that have acquired mass. How desperate can they be to invoke a spontaneous symmetry breaking mechanism?

The story continues in that prior to the symmetry breaking, the electroweak force is carried by four massless particles, W^+, W^0, W^- and B^0, for want of better names. Since they are massless they travel at the speed of light, and, like the photon, have two 'degrees of freedom' like isotopic spin. These degrees of freedom refer to the polarization of the particle in directions normal to its propagation.

The Landau-Ginsburg phase transition is now pasted on to show that the vacuum really isn't empty as we were previously led to believe. This is tantamount to someone who is changing the cards on the table once they have been dealt. In particle physics it is called the Higgs field to which the Higgs particle belongs, while in inflationary cosmology, it is referred to as the inflaton field with its associated particle. The folklore is that the massless particles, W^+ and W^- interact with the Higgs field thereby acquiring a third 'degree of freedom' — thereby becoming fat. In this way, a new direction of polarization is associated with gaining weight.

However the presence of mass has nothing to do with longitudinal waves. Nonetheless, in contrast to Maxwell's equations which require only the fields, the presence of mass requires, in addition, the potentials [Lav11b]. If mass required a longitudinal mode, it could not be polarized into the tangential and longitudinal mass, so familiar from the old days of special relativity.

The W^0 and B^0 particles 'mix together' to give a massive Z^0 particle and a photon. But it wasn't sufficient to have three massive particles, W^+, W^- and Z^0. A fourth component was also required, a boson of spin-0, the so-called Higgs boson. Lederman calls the Higgs boson the 'God particle' since it is this particle that bequeaths mass to the other massless particles. However, this just shifts the responsibility to a new particle, and doesn't answer the question of how *it* acquired mass?!

This electroweak theory, developed independently by Weinberg and Salam, was found to be renormalizable several years later by Veltman and his PhD student, 't Hooft. It is surprising that the latter received a Nobel prize for patching up a theory which, to say the least, could hardly be called fundamental. The theory made three predictions, two of which does not depend upon symmetry breaking, while the third could just be luck, if the properties of the Higgs boson live up to all its expectations. We'll just have to wait and see.[3]

---

[3]It now appears that the God particle — the Higgs boson — may actually be a *composite* particle, which would place in jeopardy not only the standard model, but, in addition,

Not much imagination was required to transform the Higgs mechanism into a cosmological context. But it still does not answer the question of what this field really is, other than an expedient way of shirking the question of how particles become massive — if that question has any real meaning to it. A much more promising approach is to return to the Proca equations which Yukawa used to help himself to a Nobel prize. What the Romanian physicist, Proca, did — which was no small feat — was to modify Maxwell's equations so they would admit a non-vanishing photon mass through the appearance of the Compton wavelength in the modified equations. That required the potentials, and not just the fields, as in the original Maxwell equations. Proca deserves more credit than he was given.

the 2013 Nobel prize in physics. According to Belyaev, Brown, Froadi and Frandsen in their paper, "Technicolor Higgs boson in the light of LHC data," published in *Phys. Rev. D* **90** 035012 on the 13th of August 2014, one specific techni-quark particle has a resonance that comes within the uncertainty of measurements for the Higgs boson. The term technicolor conveys the meaning of composite in creating particles or color. This may only go to show that cute tricks do not necessarily go hand-in-hand with physical reality!

Bibliography

[ACW74] R. L. Johnson, C. B. Thorn, A. Chodos, R. L. Jaffe and V. Weis-
skopf. A new extended model of hadrons. *Phys. Rev. D*, 9:3471–
3495, 1974.

[AM04] P. M. Alsing and P. W. Milonni. Simplified derivation of the
Hawking-Unruh temperature for an accelerated observer in vacuum. *Am. J. Phys.*, 72:1524–1529, 2004.

[Bor09] M. Born. Die Theorie des starren Elektrons in der Kinematik
des Relativitätsprinzips. *Ann. der Phys.*, 335:1–56, 1909.

[CG52] H. B. Callen and R. F. Greene. On a theorem of irreversible
thermodynamics. *Phys. Rev.*, 86:702–710, 1952.

[Dav75] P. C. W. Davies. Scalar particle production in Schwarzschild
and Rindler metrics. *J. Phys. A*, 8:609–616, 1975.

[Ful73] S. A. Fulling. Nonuniqueness of canonical field quantization in
Riemannian spacetime. *Phys. Rev. D*, 7:2850–2862, 1973.

[GR80] I. S. Gradshteyn and I. M. Ryzhik. *Table of Integrals, Series
and Products*. Academic Press, Orlando, 1980.

[Jea19] J. Jeans. *Problems in Cosmology and Stellar Dynamics*. Cambridge University Press, Cambridge, 1919.

[Kle70] M. J. Klein. *Paul Ehrenfest*. North-Holland, Amsterdam, 1970.

[Lav91] B. H. Lavenda. *Statistical Physics: A Probabilistic Approach*.
John Wiley, New York, 1991.

[Lav07] B. H. Lavenda. High temperature properties of the MIT bag
model. *J. Phys. G: Nucl. Part. Phys.*, 34:2045–2051, 2007.

[Lav11a] B. H. Lavenda. *A New Perspective on Relativity: An Odyssey in
Non-Euclidean Geometries*. World Scientific, Singapore, 2011.

[Lav11b]   B. H. Lavenda. *A New Perspective on Relativity: An Odyssey in Non-Euclidean Geometries.* World Scientific, Singapore, 2011.

[Lav13]   B. H. Lavenda. Cosmic illusions. *Journal of Modern Physics,* 7A1:7–19, 2013.

[Max04]   J. Clerk Maxwell. *Theory of Heat.* Longmans, Green and Co., London, 1904.

[Mil94]   P. W. Milonni. *The Quantum Vacuum.* Academic Press, New York, 1994.

[MOSC04]   A. Belyanin, E. Fry, M. O. Scully, V. V. Koscharovsky and F. Capasso. Enhancing acceleration radiation from ground-state atoms via cavity quantum dynamics. *Phys. Rev. Lett.,* 93:129302, 2004.

[Par77]   L. Parker. In Asymptotic Structure of Space-Time, F. P. Esposito and L. Witten, Eds. 107–226. Plenum, New York, 1977.

[Rin66]   W. Rindler. Kruskal space and the uniformly accelerating frame. *Am. J. Phys.,* 34:1174–1178, 1966.

[Sch39]   E. Schrödinger. The proper vibrations of the expanding universe. *Physica,* 6:899–912, 1939.

[She60]   C. W. Sherwin. Some recent experimental tests of the clock paradox. *Phys. Rev.,* 120:17–21, 1960.

[Smo00]   L. Smolin. *Three Roads to Quantum Gravity.* Phoenix, London, 2000.

[Sus08]   L. Susskind. *The Black Hole War: My battle with Stephen Hawking to make the world safe for quantum mechanics.* Little, Brown & Co., New York, 2008.

[Ume93]   H. Umezawa. *Advanced Field Theory.* American Institute of Physics, New York, 1993.

[Unr76]   W. G. Unruh. Notes on black hole evaporation. *Phys. Rev. D,* 14:870–892, 1976.

[Wal84]   R. M. Wald. *General Relativity.* University of Chicago Press, Chicago, 1984.

Chapter 6

Do Negative Heat Capacities Exist?

6.1 Black Hole Density of States: Exponentially Shocking

We normally think of heat capacities as extensive quantities that are necessarily positive. But the astronomers know better: since the late nineteenth century they have known that when heat is absorbed by a star, or star cluster, it will expand and cool down [Edd26].

Landau and Lifshitz [LL58] warn us that the conditions of positive heat capacity at constant volume always refer to homogeneous systems, and that an increase in volume at constant temperature is always followed by a decrease in pressure. However a body that is held together by gravitational force constitutes an inhomogeneous system. As we approach the core, the density will rise, and the heat capacity "as a whole may be less than zero so that its temperature will rise as its energy decreases." This does not contradict the fact that the heat capacity in any of its layers is positive, but we can't use the fact that the total energy is the sum of its parts since there are gravitational interactions among its parts.

It has long been conjectured, without as yet any concrete proof that, in a relativistic degenerate star which is beyond the Chandrasekhar limit, a gravithermal catastrophe will develop that will turn it into a black hole. This has been predicted by Bekenstein [Bek74], and confirmed by Hawking [Haw74], but only through wrong analogies and wrong associations. Here the problem is not that fact that the gravitational interactions have not been properly accounted for, but, rather, the assumed density of states violates the second law of thermodynamics.

The proof is simple enough to repeat here [Lav]. Consider the Unruh temperature for gravitational acceleration,

$$T = \frac{\hbar}{2\pi k} \frac{GM}{r^2},$$
(6.1)

which is to be evaluated at the Schwarzschild radius. This brings in the inverse dependency between mass and temperature, and is the root of problem. The Unruh temperature can be avoided, by simply setting the thermal wavelength, λ_T, in its ultrarelativistic expression,

$$T = \frac{\hbar c}{k \lambda_T}$$
(6.2)

equal to the Schwarzschild radius. There is no deep reason for this; rather, it is out of ignorance of any other dimension for a black hole. This will differ by a factor of 4π from expression (6.1), 2π of which is unaccounted for anyway. The upshot of the matter is that the accelerating detector measures black body radiation at a temperature determined by the Wien displacement law, $T\lambda_T = \text{const.}$, so that its 'acceleration' must be $a = c^2/\lambda_T$ [c.f., §5.6].

Using (6.1), we can solve for the inverse temperature,

$$\frac{1}{T} = 8\pi k \frac{E}{E_{Pl}^2},$$
(6.3)

where the Planck energy, $E_{Pl} = \sqrt{\hbar c^5/G}$, and the rest energy, $E = Mc^2$. The latter can hardly be considered an internal energy. Nonetheless, let's make believe that it is so that we can appeal to the second law, $dS/dE = 1/T$. Upon integration we find an absolute entropy,

$$S = 4\pi k \frac{E^2}{E_{Pl}^2},$$
(6.4)

if we agree to set the arbitrary constant of integration equal to zero. Expression (6.4) conforms to what has now become conventional wisdom that the entropy is one-quarter the area of the horizon, $S = \pi R^2/\ell_{Pl}^2$, where $\ell_{Pl} = \sqrt{G\hbar/c^3}$ is the Planck length.

Differentiating (6.3) with respect to the (rest) energy,

$$-\frac{1}{T^2} \frac{dT}{dE} = \frac{8\pi k}{E_{Pl}^2},$$

we find a negative heat capacity,

$$C = -\frac{E_{Pl}^2}{8\pi k T^2}. \tag{6.5}$$

The reason for this is not the lack of additivity due to gravitational interaction, but rather to an unacceptable density of states,

$$\varrho(E) = e^{S/k} = e^{4\pi E^2/E_{Pl}}, \tag{6.6}$$

which violates the condition that it cannot increase faster than the exponential of a power of the energy less than unity. Otherwise, the entropy will not be concave with the consequence that the heat capacity will be negative. This has disastrous consequences for estimates of the temperature. It is essential to keep in mind that *a negative heat capacity is not the result of any lack of non-additivity due to gravitational interactions, but rather, is a consequence of the unlawful density of states, (6.6).*

Suppose our black hole is divided into a number of cells, each with mass m_i, and specific heats, $c(T_i) = c \cdot f(T_i)$, each cell being initially at a different temperature, T_i, because the cells are initially separated by adiabatic walls. If these walls are replaced by diathermal ones, the cells will interact thermally, and finally will come to a uniform temperature determined by the first law [Lav09],

$$\Delta E = -c \sum_i m_i \int_{T_i}^{T_f} \frac{dT_i}{T_i} = c \left\{ \frac{M}{T_f} - \sum_i \frac{m_i}{T_i} \right\} = 0.$$

The condition is satisfied if the final common temperature, T_f, is the harmonic mean temperature,

$$\frac{1}{T_f} = \frac{1}{M} \sum_i \frac{m_i}{T_i}, \tag{6.7}$$

where the total mass,

$$M = \sum_i m_i,$$

is the sum of its constituents.

A harmonic mean temperature, (6.7), is in clear violation of the third law,

$$\lim_{T \to 0} S = 0,$$

for it would imply that the entropy would tend to infinity as the temperature tended to zero. The smallest mean temperature possible is the geometric mean, and it corresponds to maximum work done by the system.

Consider, for reasons of simplicity, two subsystems. Multiplying (6.7) through by cM results in

$$E_f = \frac{cM}{T_f} = E_1 + E_2 = \frac{cm_1}{T_1} + \frac{cm_2}{T_2},$$

which is merely a statement of energy conservation. The entropy, on the other hand, needs to increase on going from a more to a less constrained state of equilibrium. This requires

$$\Delta S = c\left\{\frac{M}{T_f^2} - \left(\frac{m_1}{T_1^2} + \frac{m_2}{T_2^2}\right)\right\} \geq 0,$$

which, when expressed in terms of energies, becomes

$$\frac{E_f^2}{M} \geq \frac{E_1^2}{m_1} + \frac{E_2^2}{m_2}. \tag{6.8}$$

At the next level of simplicity, assume that the masses are equal, $m_1 = m_2 = m$, so that $M = 2m$. This has the effect of reducing (6.8) to

$$(E_1 + E_2)^2 \geq 2(E_1^2 + E_2^2),$$

or that $2E_1 E_2 \geq E_1^2 + E_2^2$, which is clearly wrong. Q.E.D. Consequently, an entropy of the type (6.4) not only violates the second law, it moreover violates the third law!

6.2   Absence of Convection Implies Positive Heat Capacities

In thermodynamics there exists a hierarchy for determining equilibria with thermal equilibrium at the top of the list. In hydrodynamics, in contrast, there can be mechanical equilibrium without there being thermal equilibrium.

For a star to be in mechanical equilibrium, the pressure must balance gravitational attraction at any point in the star. This requires [c.f., §2.1]:

$$\frac{dp}{dr} = -\rho\frac{d\Phi}{dr}, \tag{6.9}$$

which is the condition for quasi hydrodynamic equilibrium, where Φ is the gravitational potential. The temperature will decrease from the interior of a star, but if it exceeds a certain amount, which will now be determined, convection currents will set in that will cause a mixing of the fluid so as to facilitate the equilibration of the temperatures in the different layers.

Since we are dealing with an unconstrained system, the entropy must increase with the radial distance from the center of a star. Considering $S = S(p, T)$, the condition that it increase is [LL59]

$$\frac{dS}{dr} = \left(\frac{\partial S}{\partial T}\right)_p \frac{dT}{dr} + \left(\frac{\partial S}{\partial p}\right)_T \frac{dp}{dr} = \frac{C_p}{T}\frac{dT}{dr} - \left(\frac{\partial V}{\partial T}\right)_p \frac{dp}{dr} > 0, \quad (6.10)$$

where a Maxwell relation for the Gibbs free energy has been introduced in the second equality. Therefore, in order that convection be absent, it is necessary, though not sufficient in general, that

$$\frac{dT}{dr} > -T\rho \left(\frac{\partial V}{\partial T}\right)_p \frac{d\Phi}{dr} \bigg/ C_p,$$

where we have introduced (6.9).

Specializing to an ideal gas where $\rho T = (\partial V/\partial T)_p$, and V is the specific volume, and observing that the temperature is always a decreasing function of distance, the condition for the absence of convection is

$$\left|\frac{dT}{dr}\right| < \left|\frac{d\Phi}{dr}\right| \bigg/ C_p. \quad (6.11)$$

Thus, in order that convection be absent, a necessary condition is the heat capacity at constant pressure, $C_p > 0$.

We now have to distinguish between two C_p's. The heat capacity at saturation for a two phase equilibrium is [Gug67]:

$$C_p^{sat} = T\frac{dS}{dT} = T\left[\left(\frac{\partial S}{\partial T}\right)_p - \left(\frac{\partial V}{\partial T}\right)_p \frac{dp}{dT}\right] = C_p - \left(\frac{\partial V}{\partial T}\right)_p \frac{\Delta H}{\Delta V},$$

$$(6.12)$$

where the same Maxwell relation that was used in (6.10) has been inserted into the first equality, while the Carnot-Clapeyron equation has been employed in the second. ΔH is change in the enthalpy in which there is an isothermal passage of a unit amount of substance from one phase to the other, with a corresponding change in the specific volume, ΔV. It comes as no great surprise that *whereas C_p^{sat} can be negative, the specific heat C_p is always positive*. Even for an innocuous substance as steam, its heat capacity at saturation is negative, $C_p^{sat} = -75$ J/K-mol.

What does this have to do with a gravithermal system like a star, or a star cluster? According to Guggenheim [Gug67], even the simplest system when in the presence of a gravitational field must be considered as a "continuous sequence of phases each differing infinitesimally from its neighbors."

In the first equality in (6.12), we can write $dp/dT = dp/dr \cdot dr/dT$ and avail ourselves of the condition of quasi hydrodynamic equilibrium, (6.9), to get

$$C_p^{sat} = C_p + \left(\frac{\partial V}{\partial T}\right)_p T\rho\frac{d\Phi}{dr}\frac{dr}{dT}, \tag{6.13}$$

which bears a striking resemblance to (6.10) so that the same condition for the absence of convection, (6.11), applies to the condition that $C_p^{sat} > 0$. This is completely reasonable because the boiling point where steam forms is accompanied by convection.

6.3   Does a Holographic Principle Exist?

Both Hawking and Bekenstein realized that by setting the entropy of a black hole proportional to the area of its event horizon, they were dealing with huge numbers. As Hawking [Haw] explains:

> To my great surprise I found that after a burst during the collapse there remained a steady rate of particle creation and emission. Moreover, the emission was exactly thermal with a temperature [surface gravity divided 2π]. This was just what was required to make consistent the idea that the black hole had an entropy proportional to the area of the event horizon. Moreover, it fixed the constant of proportionality to be a quarter in Planck units... This makes the unit of area 10^{-66} square centimeters so a black hole of the mass of the sun would have an entropy of the order of 10^{78}. This would reflect the enormous number of different ways in which it could be made.

The phrase "enormous number of ways" gives the connotation of the relation between entropy and the logarithm of the number of microscopic configurations that are compatible with a single macroscopic state without having the slightest clue as to what those configurations are for a black hole.

Such mind-boggling orders of magnitude for an entropy left Bekenstein [Bek81] wondering how this kind of entropy would compare with normal entropies of the garden variety. It didn't require much imagination to arrive at the conclusion that the black hole entropy should be the 'mother of all entropies', and therefore be the upper bound on all entropies. In his own words, "Is it not preposterous to think there is a common denominator in two quantities so unlike in size?"

His conclusion was "black hole entropy is revealed as matching the maximal entropy for a given mass of more ordinary systems: There is no gap in magnitude between black hole entropy and ordinary entropy." This is due to the "hitherto unnoticed upper bound" to the ratio of entropy to energy

$$\frac{S}{E} < 2\pi R, \tag{6.14}$$

where R is some 'effective' radius in Planck units where all the fundamental constants are set equal to unity. This magical radius is supposed to keep S from growing faster than E for systems with "negligible self gravity", but as a body is compressed under its own weight, R becomes of order $2E$, and the entropy "catches up" with the black hole entropy growing as E^2.

This is more wishful thinking than anything else for in order to carry out his program, Bekenstein had to bridge the gap between concave and convex functions, which is not an easy feat to say the least. It is more a bridge over troubled waters than anything else.

Before getting bogged down in the heavy machinery of a formal proof of his inequality (6.14), Bekenstein tries it out on black body radiation for which $S/E \sim 1/T$. In this example the ratio S/E can made as large as we please simply by lowering the temperature T. Since the temperature is inversely proportional to the ultra relativistic thermal wavelength, some limit has to be placed on the latter so that T should not fall too much. He makes an appeal to (nonexistent) boundary effects to "arrest the growth of S/E as T is lowered further." Here Bekenstein has to admit that this "example comes close to challenging (6.14)." But the challenge he refers to is totally illusory.

After a digression on maximizing the entropy of the canonical ensemble with respect to the constraint of constant total energy, Bekenstein comes to the conclusion that "the maximal $S/\bar{E}$ is just that β for which the partition function is unity", where $\bar{E}$ is the average energy and β is the inverse temperature.

What Bekenstein is saying is that the maximal value of the ratio is $(S/E)_{\max} = \beta_0$ for which the logarithm of the partition function is

$$\ln Z(\beta_0) = 0. \tag{6.15}$$

There is no discussion of the fact that $\ln Z$ is monotonic decreasing function of β. But what there is discussion about is one of its limits. Bekenstein correctly claims that "as $\beta \to 0$, $\ln Z \to \infty$", but incorrectly asserts that "$\ln Z \to -\infty$ as $\beta \to \infty$. Hence by continuity there exists some β for which (6.15) is satisfied and there exists a maximal $S/\bar{E}$."

This is clearly false since $S/\bar{E} > \beta$ for $\ln Z > 0$ while $S/\bar{E} < \beta$ for $\ln Z < 0$ so that there can be no maximal $S/\bar{E}$! Moreover a change in the sign of $\ln Z$ would imply a change in the sign of the Helmholtz free energy. So what was the minimum free energy for ordinary systems would become the maximum free energy for black holes!

That there is no bound on the entropy-to-energy ratio follows directly from the concavity of the entropy. We must first establish that S/E is a decreasing function. From the condition of concavity,

$$S(\lambda E_1 + (1 - \lambda)E_2) \geq \lambda S(E_1) + (1 - \lambda)S(E_2),$$

for any parameter $\lambda \in (0, 1]$, it follows that in the particular case of a slope emanating from the origin to any point on the entropy-energy curve that it cannot be inferior to the tangent of the curve at that point. Setting $S(E_2) = 0$, we get $S(\lambda E) \geq \lambda S(E)$, and dividing through by E shows that

$$\frac{S(\lambda E)}{\lambda E} \geq \frac{S(E)}{E},$$

is a decreasing function because $E > \lambda E$. Since there are no exceptions in thermodynamics, it suffices to find only one example where it fails to unravel the entire structure.

Since the entropy-to-energy ratio is decreasing,

$$E^2 \frac{d}{dE}\left(\frac{S(E)}{E}\right) = E\frac{dS}{dE} - S(E) \leq 0,$$

there is no bound on the entropy-to-energy ratio,

$$\frac{dS}{dE} \leq \frac{S(E)}{E}. \tag{6.16}$$

Any chord connecting two states on the entropy-energy curve will always lie below the curve. If one of those states happens to be absolute zero then the slope of the line through the origin will always be greater than the slope of the curve at the same point. This is what (6.16) says: the ratio of the (absolute) entropy-to-energy can never be inferior to the slope of the entropy-energy curve to the same point. This inverts the Bekenstein inequality for all systems satisfying $\ln Z < 0$, as they all must do.

Therefore, the bound on the entropy-to-energy ratio would exist only if $\ln Z$ could become less than zero. For then it would no longer be 'completely monotone', meaning the signs of successive derivatives alternate in sign as they must for generating functions, like $Z(\beta)$.

Bekenstein is, in fact, advocating writing $\ln Z = -\beta^2/16\pi$ as the partition function of a black hole; it is clearly not a completely monotone function. Unlike black body radiation where an infinite set of moments exist, black holes have only two moments, and since the second one is negative, i.e., negative energy dispersion, no probability distribution can be associated with it [LDD88]. Said differently, a system with an entropy given by (6.4) simply does not exist. And the absence of a probability distribution signifies the absence of the phenomenon it was supposedly to describe. In this respect, statistical thermodynamics is no different from other games of chance. It works when there is a sufficiently large number of (honest) players so as to justify Stirling's approximation [c.f., §5.4]:

The 'bound', which applies only to negative $\ln Z$, has become known as the Bekenstein bound which states that:

> The amount of information that can be encapsulated in any region is not only finite, it is proportional to the area of the boundary measured in Planck units. The relationship that the number of bits of information the observer can gain cannot exceed a quarter of the surface area.

The Bekenstein bound was later 'generalized' by 't Hooft into what he referred to as a holographic principle in which:

> The description of a volume of space can be thought of as encoded on a boundary of the region, preferably a light-like boundary like a gravitational horizon. The entire universe can be seen as a two dimensional information structure 'painted' on the cosmological horizon such that the three dimensions we observe are only an effective description at macroscopic scales and at low energies.

The holographic principle laid dormant until Maldacena brought it into the spotlight by proposing a correspondence between an anti de Sitter universe and conformal field theory [c.f., §7.3]. Inspired by the holographic principle, he proposed a duality between a gauge theory living on a four dimensional boundary, (3+1) dimensions, and a five dimensional region, (4+1) dimensions. The latter is dear to the hearts of string theorists because it allows reconciliation between black holes and black body radiation [c.f., §7.2]. In a space of four dimensions, the surface area becomes proportional to the volume of three dimensional space, like the cavity of black body radiation.

It would be too flattering to call this pseudo science, but it just goes to show how one wrong and ill-conceived idea proliferates into a plethora of other wrong results. And what is even more distressing is that string theory and quantum loop theory have claimed to have confirmed black hole thermodynamics. The 'proof' of black hole entropy by Vafa and Strominger "is one of the few aspects of string theory that can be cited as actively confirming the theory in a testable way, so it's rather important" [JR10]. One wrong result leads to another, it's as simple as that!

6.4 Negative Heat Capacity Conundrum

To appreciate the present state of affairs, it suffices to remark that Lynden-Bell [LB], arch supporter of negative heat capacities, faults Schrödinger [Sch48] for providing such a simple 'proof' of why heat capacities are always positive. For, according to Lynden-Bell, "it is most surprising that it could ever be WRONG."

In the canonical ensemble, the average energy is

$$\langle E \rangle = \sum_i E_i e^{-\beta E_i} \Big/ \sum_i e^{-\beta E_i},$$

where β is the inverse absolute temperature. Taking the derivative with respect to T leads to

$$C_V = \frac{d}{dT}\langle E \rangle = -k\beta^2 \frac{d}{d\beta}\langle E \rangle = \left\langle (E_i - \langle E \rangle)^2 \right\rangle \geq 0.$$

Q.E.D. Yes, Schrödinger is correct, and Lynden-Bell is wrong.

It also explains why the Bekenstein-Hawking 'entropy', (6.4), is wrong, and why Hawking's [Haw76] 'realization' that, since the canonical ensemble won't work, one is forced to use the microcanonical ensemble is total nonsense.

Not only can a temperature not be defined in the microcanonical ensemble, a canonical ensemble can also be whittled out of a microcanonical one by considering that a small part of the entire ensemble is in thermal contact with the rest. The problem is that no exponential factor, $e^{-\beta E_i}$, will ever compensate the exponential growth of the density of states, (6.6), so as to form a sharp peak about the average energy, $\langle E \rangle$.

Both Thirring [Thi70] and Lynden-Bell [LB] confuse the negative total energy of a bound system with the internal energy. Whereas the former is negative, the latter is always positive. Clausius's virial theorem states that for a non relativistic system, the sum of twice the kinetic energy $\mathcal{T}$ and the potential energy $\mathcal{V}$ vanishes, i.e.,

$$2\mathcal{T} + \mathcal{V} = 0. \tag{6.17}$$

And since the total energy is $\mathcal{E} = \mathcal{T} + \mathcal{V}$, it is negative, $\mathcal{E} = -\mathcal{T} < 0$, on the strength of the virial theorem, (6.17).

In particular, for an ideal gas, $\mathcal{T} = \frac{3}{2}NkT$, and the total energy is $\mathcal{E} = -\frac{3}{2}NkT$. Consequently, the heat capacity at constant volume is $C_V = -\frac{3}{2}Nk < 0$. But — and this is a big BUT — the total energy is *not* the internal energy, and its derivative with respect to the temperature is *not* the heat capacity. So Thirring and Lynden-Bell are both wrong.

To make his point, Thirring considers a composite system comprising of two subsystems. The subsystems can exchange energy between themselves while keeping the total energy, $E = E_1 + E_2$, constant. Their temperatures are $T_i = (\partial S_i/\partial E_i)^{-1}$, and their heat capacities, $C_i = -(\partial^2 S_i/\partial E_i^2)^{-1}/T_i^2$ have positive, $C_1 > 0$, and negative, $C_2 < 0$, components such that $C_1 + C_2 < 0$.

For small fluctuations from equilibrium, $|\delta E_1| \ll 1$, and $\delta E_1 = -\delta E_2$, the change in entropy to second order is

$$\Delta S = \left(\frac{1}{T_1} - \frac{1}{T_2} \right) \delta E_1 - \frac{1}{2} \left(\frac{1}{T_1^2 C_1} + \frac{1}{T_2^2 C_2} \right) (\delta E_1)^2 + \cdots$$

The vanishing of the first order term provides the condition of thermal equilibrium, $T_1 = T_2$: a uniform temperature. The condition for a stable equilibrium rests with the coefficient of the second order term; in order for there to be a stable equilibium the inequality,

$$\frac{\partial^2 S_1}{\partial E_1^2} + \frac{\partial^2 S_2}{\partial E_2^2} < 0, \tag{6.18}$$

must be satisfied. And since the condition, $(C_1 + C_2)/C_1 C_2 > 0$, is satisfied, where's the error?

So Thirring's theorem would read: "A negative C_V system can achieve a stable equilibrium in contact with a positive C_V system provided that their combined heat capacity is negative." Lynden-Bell [LB] tells us to:

> Imagine that the negative C_V system 'Minus' is initially a little hotter than (higher T) than the positive C_V system 'Plus'. Then heat will flow from Minus to Plus. On losing heat Minus will get hotter (i.e., its temperature increases) but on gaining heat Plus will also get hotter. However, because Plus has a lesser $|C_V|$ its temperature is more responsive to heat gain than Minus's is to heat loss. Thus Plus will gain temperature faster than Minus and a thermal equilibrium will be attained with both Plus and Minus hotter than they were to start off.

Simply by the transfer of heat from a higher to a lower temperature, bodies have more heat at the end than before the transfer began.

This immediately brings to mind perpetual motion machines of the second type, which after the completion of a cycle have energy to spare. Carnot would indeed agree because there would be more heat that could be transformed into work than before the transfer began. Moreover, thermodynamics remains moot on the rate of heat transfer so Plus's receptiveness to heat gain is mere figment of the imagination.

To further emphasize why Plus and Minus can't coexist in thermodynamic harmony, consider the transition between subsystems at initial temperatures T_1 and T_2 to a common final temperature T_f. The first law,

$$\Delta E = \sum_i \int_{T_i}^{T_f} C_i dT_i = (C_1 + C_2)T_f - C_1 T_1 - C_2 T_2 = 0,$$

establishes the weighted arithmetic mean

$$T_f = \frac{C_1 T_1 + C_2 T_2}{C_1 + C_2},$$

as the final common temperature. The heat capacities, C_i appear as weights in the average.

Thus, considering a negative heat capacity is therefore tantamount to considering a negative weight, or probability for an event T_i. The fact that the total probability, $C_1 + C_2$, is negative does not correspond to a physical state. Moreover, in order that the final common temperature be positive it will not suffice to have $|C_1| > C_2$, but it also requires the condition

$$\frac{T_1}{T_2} > \frac{C_2}{|C_1|},$$

on the initial temperatures. This means that the initial 'events', T_i are not independent, contrary to hypothesis. This is completely foreign to thermodynamics as well as to probability theory. However, it appears that the condition of the means being a monotonically increasing function of their order is not jeopardized by negative weights, just their physical meanings.

Hawking [Haw76] gives us a specific example of gravitons and black holes sharing an insulated box of volume, V, where there is a certain amount of energy, E. "Either all the energy could be in gravitons or the energy could be divided among gravitons and one or more black holes. For a given energy E_1 in gravitons, the number of microscopic configurations is a sharp maximum when the gravitons are distributed as black body radiation at temperature"

$$T_1 = \left(\frac{15E_1}{\pi^2 V}\right)^{1/4},$$

corresponding to an entropy

$$S_1 = \frac{4\pi^2 V T_1^3}{45}.$$

The entropy of the black hole is $S_2 = 4\pi E_2^2$. It is left unexplained how a number can have a sharp maximum, and what are the alternative distributions for gravitons when they are not distributed as black body radiation. Is gravitational radiation the same as thermal radiation, as Hawking supposes? Since it has never been observed we, along with Hawking, can imagine whatever we like.

The condition of thermal equilibrium requires

$$T_1 = \left(\frac{15E_1}{\pi^2 V}\right)^{1/4} = \frac{1}{8\pi E_2} = T_2, \tag{6.19}$$

while the condition of stability, (6.18), is

$$E_1 < \frac{1}{4}E_2. \tag{6.20}$$

From this Hawking concludes that:

> In order for the configuration of a black hole and gravitons to maximize the probability, the volume V of the box must be sufficiently small that the energy E_1 of the black body gravitons is less than $\frac{1}{4}$ the mass of the black hole... If this condition is satisfied, the equilibrium between the black hole and the black body radiation will be stable because, if a statistical fluctuation causes a slight excess of radiation to be absorbed by the hole, the temperature of the radiation will fall more than the black hole, so the rate of absorption will decrease more than the rate of emission.

Even without going into details we know that thermodynamics can never make numerical predictions. For that you need a specific model so, in this sense, thermodynamics is a black box. It can only provide information on ratios of rate constants, not the rate constants themselves. For that you need a specific model.

With this in mind, consider the condition for equal temperatures, (6.19); it allows us to eliminate E_2 from inequality (6.20) to give

$$E_1 T_1 < \frac{1}{32\pi}. \tag{6.21}$$

The black hole relation, $E_2 T_2 = 1/8\pi$ has miraculously transformed the same inverse energy-temperature dependency onto the black body radiation. Condition (6.21) is independent of the volume of the box so we can make it as large or small as we please. This leaves us no other option than to conclude that both (6.20) and (6.21) are nonsense. *The stability condition (6.18) can only be applied on subsystems of the same kind, and the condition for it to be satisfied is that the heat capacity at constant volume be positive.*

Bibliography

[Bek74]   J. D. Bekenstein. Generalized second law of thermodynamics in black-hole physics. *Phys. Rev. D*, 9:3292, 1974.

[Bek81]   J. D. Bekenstein. Universal upper bound on the entropy-to-energy ratio for bounded systems. *Phys. Rev. D*, 23:287–298, 1981.

[Edd26]   A.S. Eddington. *The Internal Constitution of the Stars*. Cambridge University Press, Cambridge, 1926.

[Gug67]   E. A. Guggenheim. *Thermodynamics*. North-Holland, Amsterdam, 1967.

[Haw]   S. W. Hawking. The nature of space and time. Technical Report. arXiv:hep-th/9409195.

[Haw74]   S. W. Hawking. Black hole explosions? *Nature*, 248:30–31, 1974.

[Haw76]   S. W. Hawking. Black holes and thermodynamics. *Phys. Rev. D*, 13:191–197, 1976.

[JR10]   A. Z. Jones and D. Robbins. *String Theory for Dummies*. John Wiley, Hoboken NJ, 2010.

[Lav]   B. H. Lavenda. What's wrong with black hole thermodynamics? Technical Report. arXiv:1110.5322v1.

[Lav09]   B. H. Lavenda. *A New Perspective on Thermodynamics*. Springer, New York, 2009.

[LB]   D. Lynden-Bell. Negative specific heat in astronomy, physics and chemistry. Technical Report. arXiv:9812172v1.

[LDD88]   B. H. Lavenda and J. Dunning-Davies. On the law of error for mass fluctuations in black holes. *Classical and Quantum Gravity*, 5:L149–L154, 1988.

[LL58]   L. D. Landau and E. M. Lifshitz. *Statistical Physics*. Pergamon Press, Oxford, 1958.

[LL59]   L. D. Landau and E. M. Lifshitz. *Fluid Mechanics*. Pergamon Press, Oxford, 1959.

[Sch48]   E. Schrödinger. *Statistical Thermodynamics*. Cambridge University Press, Cambridge, 1948.

[Thi70]   W. Thirring. Systems with negative specific heat. *Z. Phys.*, 235:339, 1970.

Chapter 7

Has String Theory Become a Religion?

7.1 From Hadrons to Strings

Bondi [Bon60] once remarked that "if a tool designed for one job fails to do it, but turns out to be useful in a different and unexpected job, one's faith in the tool is nevertheless to some extent weakened." String theory does not even enjoy such a status since no physical prediction using the theory has ever been made, nor is it likely that it ever will. It is for this reason that we have relegated it to the last chapter of the book since a theory that cannot be refuted is hardly a theory at all. Since we are not looking for specific flaws of the theory a more qualitative approach will be followed in this chapter to uncover the logical inconsistencies of visualizing particles as strings.

String theorists, or stringists, are full of unrestrained exaggerations of what string theory — or its super-version — accomplishes. "String theory is an excellent candidate for a unified theory of all forces of nature... String theory is a quantum theory, and, because it includes gravitation, it is a quantum theory of gravity" [Zwi09]. To say this to veteran physicists is one thing, but to preach it to undergraduates is totally unethical.

String theory can be summarized in a nutshell: Each and every particle is considered as a particular vibrational mode of an elementary microscopic string. The absence of adjustable parameters is an indication of uniqueness, and is a definite plus in string theory's favor. The number of spacetime dimensions emerges from a calculation, and, after taking into consideration fermions as well as bosons, the spacetime dimension is not four but ten! But there is consolation insofar as the presence of fermions has reduced the twenty-six dimensions if there were only bosons to a mere ten. We should be happy for the little things in life.

Rather than being independent particles, bosons and fermions are related through 'supersymmetry'. Every boson has his fermion partner. This is, indeed, surprising for what determines the statistics is spin, and there is nothing in common between the spins of bosons and fermions. Integral spin makes bosons gregarious, while half-integral spins make fermions exclusive. There is no reason for their partnership, and recent experiments point to the non-existence of supersymmetry.

Gauge type theories arise in string theory when D-branes are introduced. D-branes are extended objects, or hyperplanes, in string theory upon which strings can dock. It has even been claimed that "black holes can be built by assembling together various types of D-branes in a controlled manner" [Zwi09], whatever 'controlled' means. Even more miraculous is the fact that our four dimensional world is a D-brane inside a ten dimensional spacetime.

Just as the fifties witnessed rapid advances in the development of quantum electrodynamics, the sixties was dedicated to the unraveling of the mysteries surrounding the strong interaction. Hopes in arriving at an understanding of the physical phenomena in the strong interaction have long been dashed, and particle physicists' sights were lowered to a mere description of them, and possible predictions. Relativistic scattering was described in terms of particle exchange. And where physical laws were missing, symmetry snuck in to take their place. The property of duality was employed to express the scattering amplitude of low energy phenomena as exchange of resonances in the direct channel, while at higher energies duality was interpreted as the exchange of Regge poles in the cross channel.[1] Hadronic states were classified as either being a bound state, if the particle is stable, or a resonance for an unstable particle.

Chew and Frautschi discovered that when particles with the same internal quantum numbers, but with different spins and masses, were plotted with the spin as a function of the square of the mass, linear trajectories were obtained. These trajectories characterize a 'Regge pole', which is a

---

[1] The cross channel is obtained by switching one of the reactants and one of the products with their antiparticles and reversing their momenta so that the reactant becomes a product and the product becomes a reactant.

pole in the scattering amplitude in the complex angular momentum plane. The advantage of considering complex angular momenta was that they are analytic everywhere except at the poles where they blow up. These poles were named after Regge who introduced them into the field of strong interaction.

A field theoretic description that proved so useful in quantum electrodynamics was seemingly useless in a description of the strong interaction so that a more phenomenological approach was deemed more fruitful. A simple Lagrangian was not available into which one could feed mass and coupling values in the hope that other physical observables could be calculated by a perturbation series in terms of a small parameter, analogous to the fine-structure constant in electrodynamics. This was because the relevant coupling constant turned out to be large so that the series would converge only slowly, or not at all.

Attention was shifted to what was referred to as an S-matrix, or scattering matrix, based on the physical property of unitary. That is, the absolute square of the S-matrix should be unity, expressing conservation and maximum analyticity. It was also required that the analytic structure of the S-matrix should be as simple as possible, and satisfy self-consistency requirements, referred to as 'bootstraps', when nothing else was given.

Physical principles gave way to the search for symmetries in scattering processes. In two-body scattering, it was observed that the same scattering amplitude was found when a particle was replaced by its antiparticle. This was referred to as 'crossing' symmetry. In addition, the form of the scattering amplitude was known in the high-energy, or the so-called Regge, limit.

However, S-matrix theory suffered from its phenomenological foundations; simply said: what you put in you got out. But, it was the only game in town; that is, until 1968 when Veneziano succeeded in writing down a closed form expression of a four-particle scattering function in terms of the symmetric Euler beta function. At one end of the spectrum, where the energy tends to infinity at constant momentum, the beta function reduced to the high-energy Regge limit. At the opposite end of the spectrum, known as the hard scattering limit, where the energy tends to infinity along with the momentum in such a way that their ratio remains constant, the scattering amplitude showed a power law decay.

In this limit, the Veneziano scattering amplitude failed miserably; not only was it not unitary, meaning that the sum of all the transition probabilities was not one, but it moreover did not reflect the particle, or 'parton-like' behavior in the high energy, fixed angle, limit where the amplitude falls off as a power of the energy.

Notwithstanding the symmetry and generality of the beta function it was destined for the bin. Contrary to the assertions made in the popular scientific literature, "Veneziano's observation provided a powerful mathematical encapsulation of many features of the strong force and it launched an intense flurry of research aimed at using Euler's beta function, and its various generalizations, to describe the surfeit of data collected at various atom smashers around the world" [Gre00]. However, it wasn't the beta integral that was the issue but rather the exponents in the integrand of the integral which were angles related to the Regge trajectories.

Since many degrees of freedom are involved, it was recognized that a certain mapping function from one complex plane to another provided the key to the generalization of Veneziano's beta integral. The function maps a given polygon in one complex plane to the real axis of another plane which represents the physical poles. The angles of the polygon are what correspond to the Regge trajectories.

This type of 'conformal', or angle preserving, mapping was used in the second half of the nineteenth century to solve electrostatic problems involving sources and sinks, fluid flow, and thermal diffusion. The conformal mapping goes under the names of Schwarz and Christoffel, two eminent nineteenth century mathematicians. Applying conformal mapping to many of the electromagnetic problems found in Maxwell's treatise, J. J. Thomson was able to solve them with greater simplicity and elegance.

Early on in the study of the strong interaction it was found that linear plots resulted from plotting the spin of a particle against the square of its mass. Recently [Lav13], it has been shown that this relation represents a generalization of the Euclidean conservation of angular momentum to the non-Euclidean hyperbolic plane of negative constant curvature, like that of a saddle. Being a linear plot there are two independent parameters: the slope and intercept which could be obtained from the Chew-Frautschi plots.

Whereas the slope is almost universal to all particles, the intercept is not. To interpret the particles as tiny strings wiggling around and satisfying a wave equation meant setting the slope inversely proportional to the string tension, and an intercept of unity, the highest possible one, required no less than twenty-six dimensions! This was the condition for a complete set of harmonic oscillator states, and the elimination of 'ghosts', or states of negative probabilities. Nevertheless the lowest state in the spectrum had a vanishing angular momentum, and this meant that the square of the mass was negative. Such particles are known as tachyons which because of their imaginary mass travel faster than the speed of light. If not merely a nuisance, tachyons signaled the irrationality of string theory.

Since such particles are outlawed by the theory of special relativity, this would have been a good point to file the theory away in the bin. But not to string theorists; if it is not a theory of strong interactions it may be a theory of something else. You have to given them credit for their perseverance in the face of daunting obstacles. In this respect, it is sufficient to recall the words of Bondi in §7.1.

By setting the intercept equal to unity, string theorists were selecting out one particle in particular, the so-called "Pomeron", named after the Russian particle physicist Pomeranchuk. That particle constitutes an extreme situation where the total cross section for scattering becomes a constant, and is no longer a decaying power of the momentum transfer. This was another nail in the coffin of string theory being a theory of the strong interaction.

All this would have been sufficient for the demise of string theory were it not for the fact that "the mathematical structure of string theory was so beautiful and had so many miraculous properties that it had to be pointing to something deep," according to Schwarz, one of the pioneers and staunch supporters of the theory.

And lo and behold it came to pass that Schwarz and Scherk found that these vibrating states described a massless particle of spin 2. The particle, baptized the "graviton", has not been observed to date, but it is fervently believed to be the carrier for the gravitational field strength, just like the photon, which is a spin 1 particle, is the carrier for the electromagnetic field strength. Since carriers do not travel instantaneously fast, it asserts that gravitational waves propagate at finite speeds, and Einstein, taking the cue from Poincaré, set it equal to the speed of light. Perhaps these "dual models are instead the correct way to include particles of 'excessively high spin' in quantum gravity?" [MBGW87].

How can all particles be included in string theory when the intercept selects out one particle, albeit a limiting form? If string theory is "not just a theory of the strong force," but "is a quantum theory as well" [Gre00], then it should be a TOE—a theory of everything. But, string theory was originally abandoned because it could not describe the hard scattering limit, so how does its resurrection as a quantum theory of gravity — which is actually an oxymoron — again make it a candidate for a theory of the strong interaction? And if it is not a theory of the strong interaction why insist on Regge trajectories as the pertinent equations of motion? There is more skulduggery in string theory than meets the eye.

7.2   Black Holes, Black Bodies and Strings

According to Bekenstein, black hole entropy is proportional to the area of the event horizon and "not to the volume enclosed in the event horizon. The failure of extensitivity is a feature of gravitational physics" [Zwi09]. By replacing volume by area does not mean a loss of extensitivity, but the fact that the entropy should be proportional to the *logarithm* of the volume of phase space occupied by the system means that the density of states should increase with the exponent of the energy, and not faster. And if the density of states increases exponentially as the first power of the energy, as it supposedly does for strings, the maximum temperature will be reached, which is referred to as the Hagedorn temperature, or the temperature of the primordial fireball.

The Hagedorn temperature is the highest possible temperature, and is achieved in the limit where the entropy loses its property of concavity, i.e., heat capacities should be positive. At the Hagedorn temperature the heat capacity becomes a constant just like in the law of Dulong-Petit. This being the case we can rightly question how temperatures higher than the Hagedorn temperature can be realized, say by a black hole, where the entropy is proportional to the square of the energy, rather than being linear in the energy as it is for strings.

The final point that needs to be addressed is how can black holes radiate thermal black body radiation when black hole are confined to two dimensions while black bodies to three? This is undoubtedly the easiest to answer: Just increase the number of dimensions so that the 'area' of the black hole horizon becomes a three sphere with a finite 'volume', A_{hor}. Consequently, the five dimensional analog of the black hole entropy is

$$S_{bh} = \frac{A_{hor}}{4G^{(5)}}, \tag{7.1}$$

where $G^{(5)}$ is the five dimensional analog of Newton's constant [Zwi09], in units where the universal constants are all unity. Dimensionality is now written into what was previously a universal constant.

For relativistic strings the entropy is shown to be proportional to the square root of the the total spin. When this is large compared to an intercept which is unity, the linear Regge trajectory,

$$M^2 = \frac{1}{\alpha'}(J-1) \simeq \frac{1}{\alpha'}J, \tag{7.2}$$

shows that the entropy is proportional to the mass, and with an abuse of energy concepts, proportional to the internal energy E. Even Zwiebach finds it unusual for it leads to a constant temperature, which he identifies as the Hagedorn temperature. It means that the energy can be increased as much as we like while the temperature remains the same—the highest temperature possible.

As the temperature nears the Hagedorn temperature, $T_\star$, the energy increases without limit as

$$E = \frac{T_\star}{1 - T/T_\star}, \quad T \approx T_\star. \tag{7.3}$$

Contrary to what has been claimed [Zwi09], this is *not* a string effect, but rather represents an upper limit on the temperature when the density of states grows exponentially with the energy.

It is amusing, if hardly instructive, to see how the string theory people derive (7.3). They use a number theoretic result derived by Hardy and Ramanujan for the number of partitions, $p(N)$, of N. A partition of N is a set of positive integers whose sum is N. What is being partitioned here is immaterial, and hardly relevant to the problem. So we can easily forget this particular.

To leading term in the asymptotic expansion of the number of partitions of N, Hardy and Ramanujan find

$$p(N) = \frac{1}{4N\sqrt{3}} \exp\left(2\pi\sqrt{\frac{N}{6}}\right). \tag{7.4}$$

Since (7.4) is a big number its logarithm might be proportional to the entropy. So why not have a go and set $S = \ln p(N)$? At high energies we have a Regge trajectory without an intercept, $N = \alpha'E^2$, by confusing the number of partitions of N with the spin.

Then invoking the second law, $1/T = dS/dE$, there results [Zwi09]

$$\frac{1}{T} = 4\pi\sqrt{\alpha'} - \frac{27}{2E} = \frac{1}{T_\star} - \frac{27}{2E}. \tag{7.5}$$

As is becoming a habit, we get the unexpected result, $T > T_\star$. What, we have temperatures greater than the highest temperature possible, $T_\star$? Rather than chucking the whole analysis out the window, the stringists in the face of such negative results remain undaunted.

Stoically plodding on, they solve (7.5) for the temperature and get

$$T = \frac{T_\star}{1 - 27T_\star/2E}, \tag{7.6}$$

"which can be trusted only at high energies" [Zwi09]. Rather, (7.6) can't be trusted at all since as $E \to \infty$, T approaches the Hagedorn temperature, $T_\star$, from *above*!

Moreover that the specific heat,

$$C = \frac{dE}{dT} = -\frac{2}{27}\left(\frac{E}{T_\star}\right)^2 < 0,$$

should cause no reason for alarm because a "negative specific heat is not completely unfamiliar. A Schwarzschild black hole has a negative heat capacity: as the energy (mass) increases its temperature goes down" [Zwi09]. There seem to be black holes everywhere!

Notwithstanding the fact that the density of states of a putative black hole grows as the exponential of the *square* of the energy—and not as the energy itself as in the case of the limiting Hagedorn temperature—it is asserted that in the particular case of a five dimensional black hole its "entropy can be calculated exactly in string theory" [Zwi09].

For a five dimensional black hole, its horizon is a three sphere with a finite *volume*, A_H. In five dimensions, area has become volume and since mass is defined as density times volume the entropy is proportional to the mass, and not the square of it as for horizons which are two dimensional. But this is completely contorted reasoning since why should we associate the square of the mass with a two dimensional object when we associate the mass itself with a three dimensional one?

Disregarding this rather insignificant point, mass is energy, and the energy of black body radiation is proportional to the volume. So five dimensions allow for the leeway to make black holes compatible with black body radiation. What is miraculous is the number of physicists, if they may be called such, that have swallowed this up!

7.3 A Bridge to Cosmology: The AdS/CFT Correspondence

It is much easier to explain the CFT acronym than the AdS one. CFT stands for conformal field theory, and a field with four different supersymmetries was chosen because it is a special quantum field theory containing all massless particles. The Compton wavelength specifies a unit of length, but it is inversely proportional to the mass, so that with no mass, there is no Compton wavelength and no unit of length. This scale invariance implies conformal invariance, and it is now to be related to gauge invariance.

AdS stands for 'anti-de-Sitter', and refers to a non-Euclidean metric of negative constant curvature in contrast to the de Sitter metric which is one of positive constant curvature. The acronym sounds better than the actual wording because of all the high-flinging cosmological connotations it carries with it. The de Sitter metric belongs to spherical geometry like the metric of a sphere, only now in four dimensions with time appended onto a spherical space metric.

No matter where you are on the sphere you register the same curvature at each and every point, which depends only on the radius of the sphere. As you (stereo) graphically project the sphere onto a plane, regions closer to the poles become exaggerated in size. Contrarily, for a hyperbolic metric of constant curvature, bodies inside a disc decrease in size the further you go from the center, and since your rulers decrease in the same proportion it would seem like the boundary of the disc was infinitely far away.

This is precisely what happens in Escher's drawing of fish on a disc: Each fish is exactly the same size, but since the boundary of the disc seems infinitely far from the origin, the fish closer to the boundary appear smaller than the ones closer to the center. Or so it would appear.

But impressions can be misleading because our Euclidean rulers are not the same as the rulers belonging to the hyperbolic plane. To the inhabitants of the disc, all the fish would appear to be the same size. In other words, the size of the fish cannot be used by the inhabitants to specify their positions on the disc.

A universal error committed by cosmologists is to assume that the coefficient of the angular part of the space component of the metric represents the square of the radius. This is true only for a flat, Euclidean metric. For non-Euclidean metrics, what they consider to be the radius is really the circumference of a hyperbolic circle when multiplied by 2π [Lav11].

Now why should the string people insist that the metric be AdS? Precisely because it contains a negative term in its stereographic magnification factor. It is this factor that converts the dot product into a generalized inner product. The vanishing of this term for real values of the radial coordinate means that the horizon of the black hole has been reached, and this coincides with the boundary of Escher's drawing.

However, relativists do not commit themselves to interpreting r as a radial coordinate. So how do you know there is any boundary at all in general relativity?

And not only do stringists insist on a metric which is AdS, but they moreover claim that the metric is a five-dimensional one which they indicate by AdS_5. With the four spatial dimensions, in addition to a single time dimension, AdS_5 is claimed to be equivalent to CFT in four dimensions. This supposedly gives us an extra space dimension to play with.

The scope of the duality between AdS_5 and CFT is a ploy to up the dimensions so that the Bekenstein-Hawking entropy, which is one-quarter the area of the event horizon in $3 + 1$ dimensions becomes proportional to the cube of the radius in $4 + 1$ dimensions. Hawking tells us, and string theory confirms it, that black holes radiate with a black body spectrum. And black bodies need a cavity to establish thermodynamic equilibrium. So by transforming a two sphere into a three sphere by increasing the dimensionality of space black hole radiation becomes compatible with black body radiation. Moreover by making the entropy proportional to the mass, it allows the connection with string theory. What more could one wish for?

Anything that has entropy must also have a temperature. Temperature arises from the irregular thermal motion of the particles contained in a given volume. Where do these particles come from in a black hole? Maldacena [Mal05] asks us to imagine a stack of Escher discs. The axis of the cylinder is supposed to represent the time axis; the cylinder itself is the three dimensional AdS space, AdS_3.

Now we are told to expect some weird things to happen in such a space because you feel yourself in the pit of a gravitational well. Any object you throw will boomerang no matter how hard you throw it. "If you send a flash of light, which consists of photons moving at the maximum possible speed, it would actually reach infinity and come back to you in a finite amount of time. This can happen because an object experiences a kind of time contraction of even greater magnitude as it gets further away from you." So when it returns to you it should experience the opposite effect and the two should somehow cancel. Not according to Maldacena.

The stack of Escher discs that form the cylinder have a constant, negative curvature, and they have nothing to do with time. Each disc in the stack is not aware of the other discs above and below it. The time axis that runs parallel to the axis of the cylinder has nothing to do with time, notwithstanding the claim that "a crucial feature of anti-de-Sitter space is that it has a boundary where time is well-defined. The boundary has existed and will exist forever. An expanding universe, like ours that comes out of a big bang, does not have such a well-behaved boundary." Has *Scientific American* put itself in the same league as *Asimov's Science Fiction* journal?

The boundary of the hyperbolic disc is well-defined, but it has nothing whatsoever to do with time. No matter how long a lapse of time there is, the Escher discs will not change so that communication between discs is not possible. If this were possible, an inhabitant of one disc would be able to give his precise location. But everything looks the same to him no matter what his precise location is. Even though they don't appear to be, all fish have the same size!

7.4 Metrics Galore!

There exists a terrible confusion over which metric describes what. The Escher metric is the hyperbolic metric of constant negative curvature. It is a three dimensional hyperbolic metric with a stereographic inner product magnification that affects both the radial and angular components but to different degrees.

The simplest black hole metric is the Schwarzschild metric in which the coefficients of time and radial coordinates are inverses of one another,

$$ds^2 = -f(r)dt^2 + f^{-1}(r)dr^2 + r^2 d\Omega_3^2, \tag{7.7}$$

where

$$f(r) = 1 + \frac{r^2}{R^2} - \frac{r_0^2}{r^2}. \tag{7.8}$$

For $r \gg r_0$, it is clear that R represents the asymptotic radius. But it would not be the radius of the Escher disc since the second term in (7.8) is positive. The other "length parameter r_0 tells us that we have a black hole" [Zwi09]. The third term in (7.7) represents the metric of a three sphere S^3.

The vanishing of the coefficient, (7.8), determines the Schwarzschild radius r_+ of the AdS black hole. However the spatial part of this metric does not coincide with the hyperbolic metric of constant negative curvature, which should be

$$f^{-2}(r)dr^2 + f^{-1}(r)r^2 d\Omega_3^2.$$

Since this coincides with the well-known Beltrami metric of hyperbolic geometry of constant curvature (3.6) the spatial part of the Schwarzschild metric, at constant curvature, can't be correct. However in respect to the time component of the metric, it is correct since the term which can be factored out makes it so. It leaves the coefficient of the time component unity, and the space component precisely that of the Beltrami metric.

The Hawking temperature is the derivative of (7.8), evaluated at r_+, divided by 4π:

$$T_H = \frac{f'(r_+)}{4\pi} = \frac{1}{2\pi}\left(\frac{r_+}{R^2} + \frac{r_0^2}{r_+^3}\right). \tag{7.9}$$

For $r_+ \ll R$, the Hawking temperature is *inversely* proportional to the radius of the event horizon for small values of the radius,

$$T_H \simeq \frac{1}{2\pi r_+}, \qquad r_+ \ll R, \tag{7.10}$$

while, for larger values of the radius, the temperature becomes proportional to the radius itself,

$$T_H \simeq \frac{1}{2\pi}\frac{r_+}{R^2}, \qquad r_+ \gg R. \tag{7.11}$$

Even to stringists, which are willing to accept almost anything that will save their theory, this is unusual: For "once a black hole is large enough, its temperature grows with size. This is an important qualitative feature of black holes in AdS space" [Zwi09]. This is typical of stringists: When they can't possibly rationalize a result they claim that is a general property, and tell the reader not to spend any thought over it. However, this is a lot to chew since, Hawking found that the temperature *decreases* with increasing mass, contrary to what (7.11) claims!

One remedy, which is not very effective, is to invoke a phase transition that separates the two types of behavior. Unsurprisingly, it is known as the Page-Hawking transition [Zwi09]. But hold your horses! We know that thermodynamics thrives on monotonic behavior, and this is enshrined in the second law. Either the entropy will show the tendency to increase, or decrease. A single experiment is sufficient to determine its behavior for all systems. This is the universality of the second law: It either holds for all systems or none. For if this were not true we could go from a positive to a negative heat capacity simply by varying a parameter!

Now we have to make the black hole entropy, (7.1), compatible with the entropy of black body radiation which varies as the cube of the temperature times the volume of the radiation cavity, $S_{bb} \sim T^3 R^3$. The last term in the metric (7.7) "shows that the horizon is a three-sphere of radius, r_+, so $A_{hor} = 2\pi^2 r_+^3$" [Zwi09]. With the five dimensional gravitational constant as $G^{(5)} = \pi R^3 / 2N^2$, where N is the number of branes, the black hole entropy (7.1) becomes [Zwi09]

$$S_{bh} = \frac{2\pi^2 r_+^3}{2\pi R^3 / N^2} = \pi N^2 (2\pi R T_H)^3. \qquad (7.12)$$

Although (7.12) has the formally correct temperature and volume dependencies, it is not a first order homogeneous function as it should be. The black hole temperature T_H is the horizon temperature of a black hole which has now been transformed into a volume. And if we insist that A_{hor} is a horizon why isn't the last term in (7.12) also considered a horizon? A mere increase in the dimensionality can never render the black hole entropy, (7.1), compatible to the entropy of black body radiation.

There is no lack of imagination in string theory. It explains the extra dimension that AdS over CFT in the following way. A meson can be thought of as a quark-antiquark pair that are glued together by the color force. The color force is mediated by a gluon which is emitted by one and absorbed by the other in the pair. The binding of the quark-antiquark pair can be pictured by lines of force, just like opposite charges in electrostatics, or strings. These gluon strings lie on the boundary of the Escher disc, and their thickness is related to how much the gluons are smeared out in space. The thickness of the string can be imagined as a coordinate, the larger the coordinate the greater is its distance to the boundary of the disc. It is this 'extra' coordinate that is needed to describe the motion within the four-dimensional anti-de-Sitter spacetime.

This 'holographic' correspondence opens up a new 'horizon' (pun intended) between the quantum theory of quarks and quantum chromo-dynamics, which treats mesons and baryons as being composed of color charges bound together by messenger gluons. The holographic picture suggests that what goes on in a volume can be described on its surface. The impetus for such a correspondence comes from the event horizon of a black hole as a two-dimensional surface enclosing an unseen, or non-existent, volume.

It is often stated that because of the high dimensionality of information theory, spanned by the letters of some alphabet or the degrees of freedom of some system, the difference between the volume and its surface is almost nonexistent when the dimensionality of the space is extremely high. The conclusion that is reached is that in a hyperspace of very high dimensionality, all the volume essentially resides very near the surface. But here the dimensionality does not warrant such a conclusion and the holographic principle is completely vacuous.

The holographic principle has been implicated in the explanation of the temperature of a black hole. Temperature is a property of particles executing random thermal motions. But the temperature applies to the event horizon of the black hole. However if the holographic correspondence is invoked it shows "that a black hole is equivalent to a swarm of interacting particles on the boundary of spacetime" [Mal05]. The particles don't have to interact to contribute to the energy, but surely, the higher their energies are the greater will be their common temperature. But haven't we made a mistake? Black hole thermodynamics claims just the opposite that the temperature is a decreasing function of the mass, or equivalently of the energy.

Are we to replace the strong force by gravitational attraction since string theory is not a theory of the strong interaction? How then can string theory be a theory of all particles predicted by the standard model? What is the gravitational attraction of quark-antiquark pairs and why should it matter in comparison to the strength of the strong force? In fact, the strong interactions of particles on the boundary cannot be explained by the gravitational interaction of strings in the interior of the disc, which at most is a model of the hyperbolic plane. These are nothing more than ludicrous attempts to salvage a theory that is not worth saving.

7.5   Will the Real Universe Please Stand Up?

Physics, over the centuries, has lowered its aim. Initially it was devised to explain the world around us. With the advent of quantum theory, it played a more modest role of providing a description of the world, even though it may not seem logical to our daily senses. Now we are asked to give up even that and accept the possibility of multiverse and even parallel universes. It turns out that physics can no longer even make predictions!

One of the Achilles' heel of inflation is what starts and stops it, while that of string theory is what is the lid on how many dimensions. Feynman opines that "string theorists don't make predictions, they make excuses" [Woi07]. For example, says Feynman:

> The theory requires ten dimensions. Well, maybe there's a way of wrapping up six of the dimensions. Yes, that's possible mathematically, but why not seven? When they write their equation, the *equation* should decide how many of these things should get wrapped up, not the desire to agree with experiment.

Not only that, but we may ask how many ways can these six extra dimensions be 'compactified', or rolled up? According to latest estimates on the number of Calabi-Yau shapes, there are not less than 100^{500} different types of possible universes. That is, some prescription must be given on how these six dimensions are to be rolled up. Imposing supersymmetry and the condition that the transformations of the two-dimensional string world sheet not be conformal, points in the six-dimensional space must have three complex coordinates and the curvature be such that a topological invariant vanishes. Calabi conjectured this in 1975 and it took another twenty years for Yau to prove it. Over the years the number of Calabi-Yau spaces have proliferated.

In the hole expansion of the two-dimensional world sheets on which strings move, there may be an infinity of these ten-dimensional spaces. The vacuum degeneracy enigma is that any one of these Calabi-Yau spaces is equally as good as any other in which to perform the superstring expansion in terms of the number of holes, which topologists refer to as "genus".

When interpreted in terms of eternal inflation, the concept eternal inflation "was greatly boosted by the realization that string theory has no preferred vacuum" [Gut07]. Whereas inflation has countless bubbles inflating in spacetime, superstring theory has colliding branes. D-branes are supposedly extended string-like objects on which different universes can live with different physical laws and different universal constants, not to mention the different types of particles which can be accommodated. Take your pick on which scenario you prefer for this is nothing less than the blind leading the blind!

String theory is actually a religion with Witten as its false prophet. His bible, *M-Theory*, explaining what superstring theory is, still has to be written. (With M remaining undefined, the readers await a Miracle, or Messiah.)

According to Woit:

> Witten's strong support for superstring theory is in itself a good reason to work in this area. Witten's genius and accomplishments are undeniable, and I feel that this is far and away the best argument in superstring theory's favor, but it is a good idea to keep in mind the story of an earlier genius who held the same position as Witten at the Institute of Advanced Study.

In fact, Witten's legacy could very well be the repeat of the domination of a cult figure in physics, with all its negative aspects.

String theorists have reason to believe that superstring theory can marry gravity and quantum mechanics that would take Einstein's classical general relativity to the next level. The term 'quantum gravity' is nothing other than an oxymoron: Gravity applies to the very large, quantum theory to the very small. Without a real quantum theory of gravity, Hawking's mixture of quantum field theory and Bekenstein thermodynamics appear as oddities at best, and pure unadulterated nonsense at worst.

If superstring theory does, in fact, describe black holes then it should be possible to calculate the entropy of a black hole and compare it with the 'classical' expression found by Bekenstein. This is precisely what Strominger and Vafa [SV96] did for an 'extremal' black hole in which there is an equivalence between mass and charge. Such an extremal black hole is a fluke for "it is an object with entropy, but no heat or temperature." One could argue the significance of the derivation of an incorrect result. But to say that there is entropy but no heat and to say that there is entropy but no temperature is to deny the second and third laws of thermodynamics, respectively.

Bibliography

[Bon60]   H. Bondi. *Cosmology*. Cambridge University Press, Cambridge, 2nd edition, 1960.

[Gre00]   Brian Greene. *The Elegant Universe*. Vintage, New York, 2000.

[Gut07]   A. H. Guth. *Eternal inflation and its implications*. Technical Report. arXiv:hep-th/0702178, 2007.

[Lav11]   B. H. Lavenda. *A New Perspective on Relativity: An Odyssey in Non-Euclidean Geometries*. World Scientific, Singapore, 2011.

[Lav13]   B. H. Lavenda. Derivation of Regge trajectories from the conservation of angular momentum in hyperbolic space. *The Open Particle and Nuclear Physics Journal*, 6:4–9, 2013.

[Mal05]   J. Maldacena. The illusion of gravity. *Scientific American*, pages 57–63, Nov. 2005.

[MBGW87]   J. H. Schwarz M. B. Green and E. Witten. *Superstring Theory*, Vol. I. Cambridge University Press, 1987.

[SV96]   A. Strominger and C. Vafa. Microscopic origins of the Bekenstein-Hawking entropy. *Phys. Lett. B*, 379:99–104, 1996.

[Woi07]   P. Woit. *Not Even Wrong*. Vintage, London, 2007.

[Zwi09]   B. Zwiebach. *A First Course in String Theory*. Cambridge University Press, Cambridge, 2nd edition, 2009.

Epilogue

1. $B-$Mode or Dust?

At the time of publication, the earth, or should I say cosmological, shattering announcement referred to in the Preface that evidence of primordial gravity waves found by the BICEP2 team last March has been retracted. Emphasis has been shifted to the precision measurements of the polarization of dust in our galaxy [Rod14] — if this is any consolation.

Had these elusive $B-$modes been found, they would have been used as evidence for primordial gravity waves that were supposedly generated by cosmic inflation, and that they would make their presence known by polarizing thermal radiation in a particular swirling pattern.

Shortly after the announcement it was realized that dust from the Milky Way would scatter light with a similar pattern. A map of the foreground polarization coming from the Milky Way is shown in Figure 1. We raised this possibility independently — and prior to its confirmation — in § 1.2.

In order to distinguish dust from the primordial $B-$modes, the BICEP2 team used the map shown in Figure 2 of galactic dust based on observations from the European Space Agency's Planck satellite, which observed the CMB from 2009 to 2013. This was supposedly preliminary information, "which was never meant to be used to analyze data from the south pole region because in that part of the sky the signal was not strong enough relative to noise" [Cow14].

The polarized fraction of radiation, $p = \sqrt{Q^2 + U^2}/I$, of total intensity I, in the upper right-hand corner of Figure 2, indicates that we are dealing with linear polarization of light, since Q and U are the Stokes parameters for the degree of polarization and the oblique angle of the linear polarization.

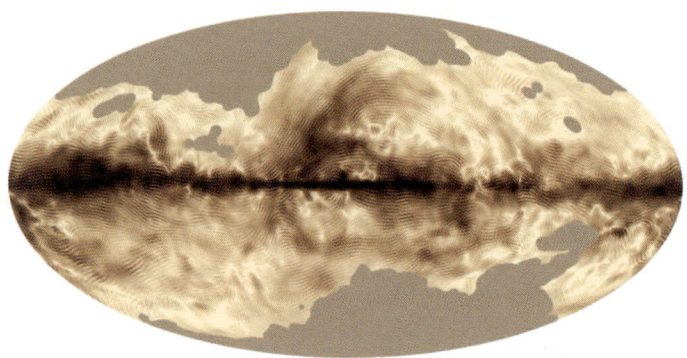

Figure 1. A map of the foreground polarization from the Milky Way. (Credit ESA and Planck collaboration)

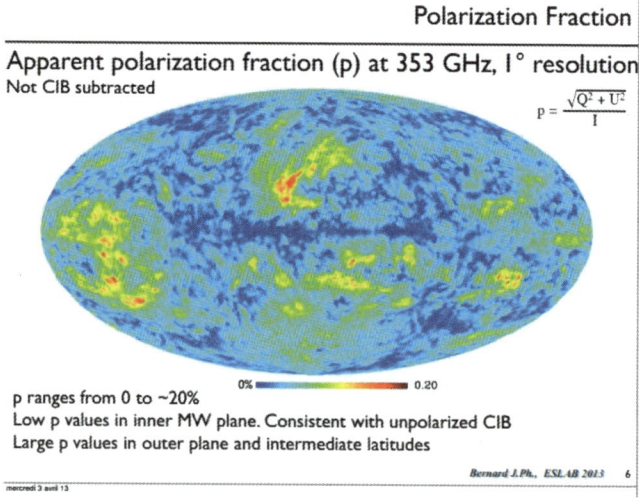

Figure 2. The imaged used by the BICEP2 team from a Planck presentation to reconstruct the data behind it.

This makes Rayleigh's law is applicable: When light is scattered normally to the direction of incidence it is plane polarized, and *this is independent of the state of polarization of the incident light* [Cha50].

The labels in Figure 2 indicate that there are two sources of radiation: cosmic infrared background radiation (CIB) and galactic dust. The CIB is a mysterious infrared light coming from outer space, which, in some ways is analogous to the CMB but at shorter wavelength. The full power of the CIB is thought to be a maximum of 10% of the CMB.

The BICEP2 team used this map to rule out dust and scynchroton radiation at a statistical significance level of 2σ. Not only were their estimates too low, but that the method they used to check their measured signal against dust was not robust [Com14]. Still another defect of BICEP2 was that its measurements were made at a single frequency of 150 GHz, instead of being made over a range of frequencies like that of the Planck probe.

So the putative 'smoking gun for inflation', or the $B−$mode detected by BICEP2, is hardly smoke, but rather dust in our own galaxy. According to the latest news report [Ach14]:

> The Planck satellite has detected dust, or more precisely the effects of galactic dust, at levels that could potentially explain away the entirety of the alleged cosmological signal reported by BICEP2 scientists at their big March 17 news conference at Harvard.

Transforming a negative result into a positive one, the leader of the BICEP2 team, Kovac, is now praising his team on the very high sensitivity measurements of the dust they made, and their aim "to take the same approach and apply it to a wider swath of sky — hopefully the whole sky" [Rod14].

Figure 3 shows the northern (left) and southern (right) projections of the sky seen by Planck. The colors indicate the amount of '$B−$mode' generated

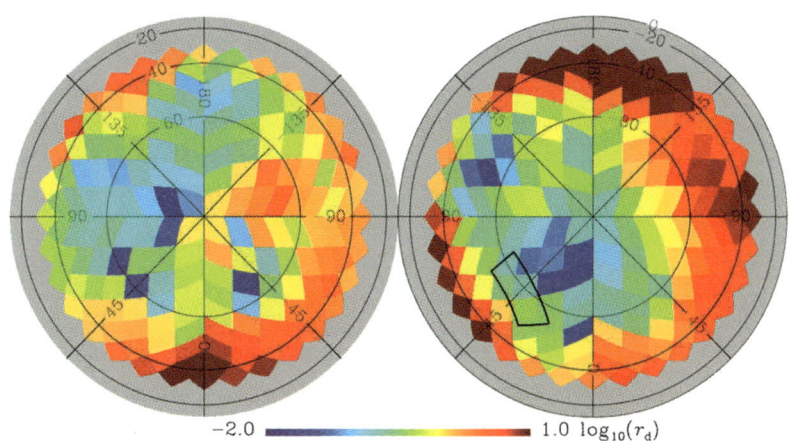

Figure 3. Planck's northern (left) and southern (right) sky map showing the projected dust contamination at 150 GHz, extrapolated from the 353 GHz data. Cleanest regions are shown in blue; dirtiest regions in red. The black contour is the approximate region that was observed by BICEP2.

by dust with the bluer areas indicating the more dust free areas. The black box on the right shows the southern projection that BICEPS2 looked at, which clearly shows that it was dust contaminated.

How were these imprints of the Almighty interpreted by the inflationists? They concocted an explanation whereby [HW97]:

> Thomson scattering of temperature anisotropies on the last scattering surface generates a linear polarization on the sky that can be simply read off from their quadrupole moments. These, in turn, correspond directly to fundamental scalar (compressional), vector (vortical), and tensor (gravitational wave) modes of cosmological perturbations.

These patterns can be distinguished geometrically in terms of the $E-$ and $B-$modes that they generate. Moreover, "polarization provides unique information for the phenomenological reconstruction of the cosmological model." The linearity of the Stokes parameters rules out such a possibility, as we shall soon appreciate.

Never has physics been so badly twisted and distorted to make the end equal the means. May we refer the reader to the principle of Optical Equivalence whereby it is impossible by means of any instrument to distinguish between various incoherent sums of simple waves that may together form a beam with the same Stokes parameters? Furthermore, the polarization of light involves the vector nature of light, so that linear polarization cannot, no less "simply", be "read off from their quadrupole moments."

Equation (1.22), therefore, has absolutely no meaning: oblique polarization, U, exists independently of the presence of a quadrupole moment, and its construction does not involve integrating over a solid angle since it will, in general, be a function of the polar coordinates ϑ and φ. It has been well-known since the time of Poincaré that the Stokes parameters depend on the coordinate system, and integrating over a solid angle would make them independent of any coordinate system.

2. Fairy Tales

The polarization of the CMB begins at recombination when photons last scattered off of electrons. Due to the ebb and flow of gravity acting to condense and radiation pressure of the photons trying to resist the condensation, temperature inhomogeneities formed where higher temperatures were found at points of compression and lower temperatures at points of rarefaction.

When recombination proceeded, photons emanating from hot and cold regions, orthogonally situated about a sole electron just waiting for the right geometry to be established, would be linearly polarized by quadrupole radiation. This is to say that photons hitting a sole electron from a given direction must be hotter than photons coming from the perpendicular directions so that the radiation emitted by the electron through Thomson scattering becomes linearly polarized along the hotter radiation axis.

In other words, an electron on the last scattering surface viewing an anisotropic distribution will generate a polarized radiation distribution if there is a non-zero quadrupole moment to the anisotropic distribution [Kos99]. Consequently, *the quadrupole anisotropy would be responsible for the polarization of the CMB.*

Are we supposed to believe that an electron is just going to sit around waiting for the right combination of photons from different directions to occur? This would give electrons at different locations different polarization orientations and different magnitudes. But since photons could not *diffuse* too far, the polarization doesn't vary much across very large angular scales.

A lot or little doesn't make a difference, photons are neither hot nor cold, and they certainly don't diffuse. A photon is defined by its frequency, and a temperature can be defined by the collectivity, not by the individual photons.

What has Thomson scattering to do with the quadrupole anisotropy? Thomson scattering applies to the scattering of *free* electrons, just like Rayleigh scattering applies to the scattering of light in the sky by free dust particles [Cha50]. But, unlike Rayleigh scattering, where the dust particle can have an internal structure, and so possess multiple moments, an electron is a point particle having no internal structure so that recourse to dipole and quadrupole moments is inapplicable. Why then introduce a quadrupole moment for the scattering of radiation by free particles?

Consider a quadrupole moment of angular momentum, ℓ, with $2\ell + 1$ different orientations, m. The scalar perturbations ($m = 0$) consist of energy density fluctuations in the primordial plasma, resulting in hotter and colder regions, and causing the velocity distribution to become out of phase with the acoustic density mode. The fluid velocity from hotter to colder regions causes a blueshift in the photon frequencies that result in a quadrupole anisotropy. The vector perturbations ($m = \pm 1$) are due to vortices in the plasma and cause Doppler shifts resulting in quadrupole lobes. However, vortex motion should be readily damped out by inflation so that its effect is supposedly negligible [HD02]. Finally, there are tensor perturbations

($m = \pm 2$) which manifest themselves in the form of gravity waves that stretch, twist and squeeze the plasma. They also increase the wavelength of the radiation thereby creating a quadrupole variation in the incoming temperature distribution [HD02].

To study the polarization pattern in the sky, it is convenient to decompose it into two components (which rather than being orthogonal are at $\pi/4$ degrees apart): A curl-free component or 'E−mode' (the E giving the connotation of electric-like), or 'gradient-mode' with no handedness, and a grad-free component or 'B−mode' (the B standing for magnetic-like) or 'curl-mode' with handedness. Whereas the E−mode may be due to both scalar and tensor perturbations, the B−mode, since it has handedness, can be due to only vector and tensor perturbations. It is the latter mode that BICEP2 claims to have picked up, and is pictured in Figure 4.

Apart from the obvious fact that the electric and magnetic fields are orthogonal to one another in free space, a second rank tensor, representing alignment and being expressed in terms of quadratic combinations of the angular momentum components cannot be decomposed into scalar, vector and tensor 'perturbations', in turn, by selecting the value of the projection of the angular momentum upon the z−axis, to be $m = 0, \pm 1, \pm 2$ [HW97].

It is well-known [BS62] that a body with angular momentum, ℓ (which can either stand for orbital or spin angular momentum), can have its angular

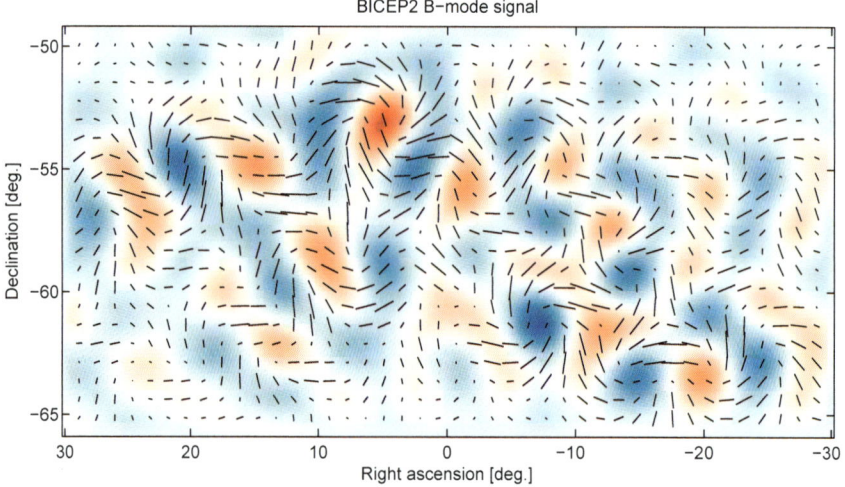

Figure 4. B−mode polarization in the CMB indicative of gravitational waves.

momentum oriented in any one of $(2\ell + 1)$ different ways, corresponding to $m = -\ell, -\ell + 1, \ldots, \ell$, where m is the projection of the angular momentum on the $z-$axis. For the quadrupole $(\ell = 2)$ there are five such orientations. If we choose the axis of the body, in this case a spheroid, so that it coincides with the $z-$axis, the only surviving member is $(3\ell_z^2 - \ell^2)$, which represents the net spin in the $z-$direction in the case of an assembly of spins. Physically, this implies an axis of cylindrical symmetry.

Such an interpretation does not tally at all with the above interpretation of scalar, vector and tensor perturbations. From this we can safely conclude that *there is absolutely no meaning to expanding the incident beam in a partial wave expansion,*

$$I' = \sum_{\ell,m} a_{\ell m} Y_{\ell m}(\vartheta, \varphi), \tag{1.1}$$

as was done in the equation following (1.21), and truncating at $\ell = 2$, the quadrupole term to determine the Stokes parameters.

All five components specify the second-rank tensor, and are required only for spin-1 particles. Indeed, for particles with spin $\ell > \frac{1}{2}$, the specification of the polarization vector is insufficient to characterize the beam [RT67]. The average polarization vector just specifies three of the quantities out of a total of $(2\ell+1)^2$ quantities. For spin-1 particles, nine components are needed: a monopole, vector, and tensor; that is, $1 + 3 + 5 = 9$ components. The five components of the second-rank tensor cannot further be reduced to three components of a first-rank tensor with $m = \pm 1, 0$, and a zero-rank tensor with $m = 0$ [HW97].

But the point is that you don't need a quadrupole moment to specify the polarization of light. The tensor components of a tensor of rank two cannot be related to the Stokes parameters. And, moreover, by a judicious choice of axes, only the spherical harmonic, $Y_{20}(\vartheta, \varphi)$, subsists. What then happens to the vector and tensor perturbations that are proportional to $Y_{2\pm 1}$ and $Y_{2\pm 2}$, respectively? And what do these have to do with $\ell = 0$ and $\ell = 1$, monopole and dipole, tensors of rank zero and one?

The dipolar pattern, $Y_{1\pm 1}$ is usually attributed to the motion of the earth and not to $Y_{2\pm 1}$, which supposedly is a vector perturbation in which "the vorticity of the plasma causes Doppler shifts resulting in quadrupole lobes" [HW97]. The Doppler effect does not, however, cause polarization!

Elementary errors have been committed throughout the literature, and also in student tutorials [HW97]. In particular, the quadrupole moment does not imply "pure $Q-$field on the sky" with Stokes parameters $Q = \sin^2 \vartheta$

and $U = 0$ [HW97]. Recall that the Stokes parameter Q is a measure of the excess of vertical to horizontal polarization, or the degree of polarization.

Since we are viewing the sky above us, which appears as a hemisphere, it is convenient to perform a partial wave expansion in terms of spherical harmonics, $Y_{\ell m}(\vartheta, \varphi)e^{im\varphi}$, of a plane wave. However, to apply the same expansion to the incident intensity and then to average over a solid angle can never yield the Stokes parameters for they would be coordinate independent whereas we know them to be otherwise.

In addition, the transformation from incoming to outgoing Stokes parameters is *linear* so that the same outgoing Stokes parameters can be due to any number of incoming Stokes parameters. There is nothing further from the truth than the statement: "The problem of understanding the polarization pattern of the CMB thus reduces to understanding the quadrupolar temperature fluctuations at *last* scattering" [HW97]. This is even more surprising when it is realized that the polarization signal is anywhere from ten to fifty times weaker than the temperature fluctuations themselves.

In fact, the Stokes parameters in Figure 2 would imply a $B-$mode according to the inflationists since, according to them a $B-$mode necessarily requires oblique, linear polarization, $U \neq 0$. It supposedly represents the twisting and shearing effects that occur when gravitational waves pass. This is completely inaccurate as Rayleigh scattering shows. Sunlight scattered by one particle becomes partially linearly polarized. This light then becomes incident for a second particle, and if the second scattering plane is not parallel to the first, the scattered light can become obliquely polarized, $U \neq 0$. No gravitational waves are involved!

A crucial test for distinguishing between $B-$modes and polarization by dust, or water droplets, would be whether circular polarization, V, was absent or not, respectively. To obtain circularly polarized light from incident linearly polarized light requires oblique polarization. But if oblique polarization occurs what would transform it into elliptical or circular polarization? There would need to be *multiple scattering* in which light scattered by one particle becomes partial linearly polarized light. Then a second scattering could create elliptical polarization if the scattering plane is rotated through an azimuthal angle, φ, with respect to the incident plane of polarization. Inflationists are adamant on the fact that simple Thomson scattering of free electrons does not produce circular polarization. If oblique polarization, $U \neq 0$, creates gravity waves, what on earth (or elsewhere) would circular polarization, $V \neq 0$ be responsible for?

3. Quadrupole Polarization

According to the inflationists, "if Thomson scattering is rapid, the randomization of photon directions that results destroys any quadrupole anisotropy" [HW97]. Supposedly, Thomson scattering of the quadrupole anisotropies generates linear polarization in the CMB by passing only one component of the polarization of the incident beam along the hot radiation axis.

Even if there were a quadrupole anisotropy, light would not pick it up. In other words, light is impervious to what type of device is used to polarize it. So why insist on a quadrupole anisotropy? Although photons are particles of spin-1, they behave more like particles of spin-$\frac{1}{2}$. In fact, the polarization of a plane light wave provides a classic example of a 2-level system. And only for spin-$\frac{1}{2}$ systems does the term 'polarization' have a unique meaning [BS62].

In general, if a spin-1 particle has a well-defined momentum, the components of its spin along the three directions of motion can take three values, $m = \pm 1, 0$. However, as a consequence of the transverse nature of electromagnetic waves, the value $m = 0$ must be omitted. Consequently, for light the specification of the polarization is sufficient for its characterization, and with it, the three Stokes parameters, Q, U, and V.

Polarization can either refer to the nature of the source of the beam or to the interaction between the beam and target [RT67]. When the inflationists refer to reading off linear polarization by the quadrupole components that it forms, they are undoubtedly making reference to the latter form of polarization. However, the former type is much more interesting since it may be a way that light can possibly account for gravity waves by adding mass to the photon. In other words, mass is the source of gravity, and if light is to be able to 'see' gravity waves, then there is no other way than to allow the photon to become heavy. We consider this possibility first.

It is only necessary to consider quantities of higher rank than the polarization vector for particles of spin $\ell > \frac{1}{2}$, except for light for the reasons just mentioned. With the spin state $m = 0$ included, it is not possible to distinguish between states with equal populations in states $m = 1$ and $m = -1$, and that with all particles in the $m = 0$ state. This state can be thought of classically as being perpendicular to the axis along the spin direction and precessing about it.

It is therefore not possible to specify a magnitude along this axis, making it necessary to consider quantities of higher rank than one [RT67]. These quantities cannot depend on the direction of the spin axis, and, hence, must be at least quadratic in the spin operators, like the components of the second-rank tensor. *Hence, they have nothing whatsoever to do with the Stokes parameters, which comprise the polarization vector.*

Quadrupole and higher moments have no meaning for particles of spin $\ell \leq \frac{1}{2}$ like electrons. This is true also of photons, as we have just mentioned, because they have only two states of polarization. What would happen if the vector wave were to acquire a 'third degree of freedom'? This question was first posed by the Romanian physicist, Proca, by generalizing Maxwell's equations to include mass, and subsequently elaborated upon by Schrödinger and his student, Bass, roughly two decades later [SB53].[1]

The introduction of a third state of polarization is tantamount to introducing mass, and with it, the previously non-existent longitudinal mode of polarization. What was a previously a pure transverse wave has now been transformed into an elastic wave with both transverse and longitudinal components. The invariants of the electromagnetic tensor no longer vanish, and the fields can be transformed to rest. The field equations require both the electromagnetic fields as well as the scalar and vector potentials for a complete characterization.

According to what has been said above, the polarization vector would be insufficient to describe the state of polarization, since it would also require a tensor of rank two. The quadrupole formed from double-sided systems is what is thought to create a gravitational disturbance resulting in gravity waves. But these waves are believed to travel as a transverse vibration far from its sources so that it has nothing to do with the third state of polarization. It would also travel at the speed of light according to general relativity. And since it is believed to have two independent states of polarization, like the electromagnetic field, there would be no need of a higher-rank tensor than the polarization vector to determine completely its state of polarization.

Thus, the general theory of relativity is plagued by fundamental incompatibilities when it attempts to characterize the polarization and propagation of gravitational waves. It cannot be of any assistance in the detection of gravity waves, and neither can the inflationary scenario since, while adhering formally to Einstein's equations, does not hold to a limiting velocity when it comes to the rate of expansion of space itself. Are we then back to

---

[1] For an historical survery and more complete discussion see [Lav11].

Laplace's idea that gravity propagates much faster than the speed of light and has characteristics profoundly different from that of light [Fla96]?

Now consider the second possibility where not only the incident beam need be taken into account, but, in addition, the target. It is claimed that CMBR can become polarized through a quadrupole anisotropy. The need for a quadrupole is clear if gravitational waves are to be accounted for. Radiation hitting an electron must be hotter in one direction than the other two normal directions. This supposedly produces linear polarization along the hot radiation axis where it is subsequently broken down into its tensoral components. The decomposition of a tensor of first rank into the components of a tensor of second rank is, indeed, a marvelous feat [HD02, HW97].

Moreover, it is claimed that "any polarization on the sky can be separated into 'electric' $(E-)$ and 'magnetic' $(B-)$ modes. This decomposition is both useful observational and theoretically..." [HW97]. However, the main advantage of the Stokes parameters is that they are measurable quantities, so why introduce new and superfluous quantities that are linear combinations of them?

The Stokes parameters are determined at each point on the wavefront by a camera placed in the observation plane together with a polarizer placed at different orientations, and, if need be, a quarter-wave, plate, or 90° retarder, if circular polarization is to be analyzed [Shu62]. The parameters are then found in a six step procedure from measured values of the irradiance for different configurations of the polarizer, in the case of linear polarization.

Of the three Stokes parameters, Q, the degree of polarization, U, the plane of polarization, and V, the ellipticity of light, V is the only invariant under a rotation in the azimuthal plane. When Q and U undergo a rotation through an angle φ in the clockwise direction, they transform into

$$Q' = Q\cos 2\varphi + U\sin 2\varphi, \tag{1.2}$$

and

$$U' = -Q\sin 2\varphi + U\cos 2\varphi. \tag{1.3}$$

It is now claimed [SZ98] that two rotationally invariant quantities, E and B, can be obtained by first taking the Fourier transform of (1.2) and (1.3) to obtain their Fourier coefficients, and *then* performing the inverse transform to "obtain them in real space" [SZ98]. Now information is not lost in performing a Fourier transform so that performing the Fourier transform, and then its inverse must give back the original quantities, (1.2) and (1.3)!

Nothing has been lost, but, then again, nothing has been gained so all this has been a useless exercise.

To say that the final transformed quantities, $E(\varphi)$ and $B(\varphi)$ are "scalar and pseudoscalar polarization fields", and that they are "two rotationally invariant fields" is pure fantasy. And this is what is precisely confirmed in the textbook [Dod03], where Eqn (10.87) are identical to (1.2), with E replacing Q', and (1.3) where B stands in for U'.

4. Building Castles in the Sky

The story of the primordial universe makes for fascinating reading if hardly scientific facts. The difference between $E-$ and $B-$modes can be taken as a measure of the cosmological signal as compared to the foreground and/or instrumental offset. $B-$mode corresponds to U polarization. Polarization can provide a clear demonstration of the existence of acoustic oscillations in the early universe. The CMB can become polarized if the radiation scattering on electrons at the moment of decoupling has a quadrupole anisotropy. Photons from a given direction hitting an electron must be 'hotter' than those hitting the same electron in the perpendicular directions. When this condition is meet, linear polarization occurs along the hot photon axis.

Gravitational waves from inflation *imprint* a unique pattern on the CMB polarization, and the anisotropies due to them produce $B-$mode polarization. Radiation gets polarized by the oscillations in the primordial plasma, and the spectrum of the polarized radiation is shifted with respect to the temperature fluctuation spectrum, where peaks and troughs are transformed into one another. Moreover,

> Gravity waves are pulsations in the metric; these induce an azimuthal dependence to the photon distribution... This dependence on φ means that there is an additional direction to choose from when the polarization field gets induced. We might expect then that the orientation of the polarization will not necessarily be aligned with the direction of the changing polarization strength. That is, we might expect that gravity waves will produce $B-$mode polarization" [Dod03].

If the $B-$mode can be caused by dust, as is now believed [Ach14], it doesn't say much for the anisotropies caused by gravity waves. If the $B-$mode *does* correspond to U polarization, who needs the $B-$mode?

Longitudinal waves, like acoustic waves, can modulate the amplitude and phase of light waves, diffract them, shift their frequency, or focus them [Adl], but they cannot change their basic constitution, like their state

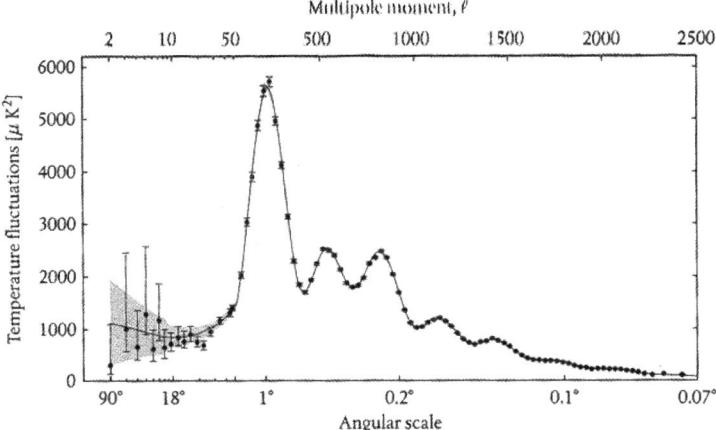

Figure 5. The structure of temperature ripples in the CMB on an angular scale, as adapted in [GS13].

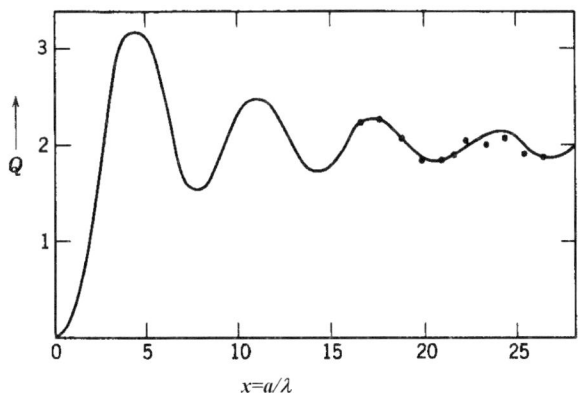

Figure 6. A typical extinction curve for dust particles.

of polarization. In other words, acoustic waves are longitudinal, and have no sectional pattern at all so they are never polarized and cannot influence or affect the polarization of light, which is transversal. They exercise different degrees of freedom which can neither affect nor interfere with one another.

There is an uncanny resemblance between the graph of the temperature fluctuation, or the intensity of the fluctuation, versus the multipole moment, ℓ, in Figure 5, and the extinction curve as a function of the size to wavelength ratio, x, in Figure 6. If the maximum value of the magnitude of the angular momentum, $|\ell|_{\max} = |rp|$ is normalized to units of Planck's constant, h, and the de Broglie relation is introduced for the linear momentum, $p = h/\lambda$, then the two coincide if the distance, r, is interpreted as the particle size, a.

Although Thomson scattering by free electrons agrees with Rayleigh scattering in the prediction of the angular distribution and state of polarization of the scattered radiation, the latter is richer in physics since there can be internal structure to dust particles where none is known for electrons. Besides electric dipole radiation there can be both magnetic dipole and quadrupole radiation [vdH57].

It all depends on the speed of light inside the particle, c/η, where η is the index of refraction. If $\eta = \infty$, the body is a perfect conductor, and the internal field is zero. That is, we assume that the fields do not penetrate the particle at all. Under this condition Rayleigh and Thomson scattering are equivalent.

For large η, there is total reflection on the surface of the dust particle, and this is the condition for the building up of standing waves, which are responsible for the optical resonance effects that appear as bumps in the extinction curves, such as those in Figure 6. Extinction, caused by absorption and scattering of light by dust particles or droplets of water, is what diminishes the intensity of the incident beam and is a function of x, the ratio of the particle size, a, to the wavelength, λ, of the radiation employed. Optical resonances appear as bumps in the extinction curve that occur at small x, and large η. As η decreases, or x increases, there is a deterioration of the resonance pheomenon.

Surface charges induce an electric dipole, while surface currents create a magnetic dipole. These appear as bumps in the extinction curve in Figure 6. The strongest resonance peak is due to magnetic dipole radiation, while the electric dipole resonance occurs for larger x, and has a smaller intensity.

In contrast, the mottled structure of the CMB has a chacteristic size, and the graph of the ripple intensity versus ripple width is shown in Figure 5.[2] Because the ripples are observed on a spherical sky, the ripple width becomes an angle, and the intensity is expressed in terms of the spherical harmonics, $Y_{\ell m}(\vartheta, \varphi)$, as in (1.1). The smaller the structure, the larger the ℓ. However, this does not mean that we can express polarization in terms of spherical harmonics, and doing so results in the wrong conclusion that, to

---

[2]The ordinate is usually expressed as 'temperature fluctuations $[\mu K^2]$', as can be found in almost any book, even those that purport experimental rigor [GS13].

lowest order, a quadrupole gives rise to the degree of linear polarization, Q, as (1.22) so definitely asserts.

By performing a partial wave expansion of the incident intensity of radiation, (1.1), it is claimed that Thomson scattering of radiation will generate linear polarization,

$$U - iV = \frac{3\sigma_T}{4\pi A} \sqrt{\frac{2\pi}{15}} a_{22}, \tag{1.4}$$

if the coefficient a_{22}, of the quadrupole moment is nonzero, where σ_T/A is the ratio of the Thomson and the geometrical cross-sections. As emphasized previously, the Stokes parameters are coordinate dependent whereas (1.4) is coordinate independent so it can't be right. Also note that the quadrupole moment is $\ell = 2$ in the partial wave expansion (1.1) so that the single coefficient a_{22} does not account for the entire quadrupole interaction of the incoming radiation. And by aligning the axis of the body with that of the field only the a_{20} will survive. This casts even greater doubts of the validity of (1.4). Finally, we take note of the fact that the coefficients in (1.1) are real, and so too are the Stokes parameters, I, Q, U and Q so that no other conclusion can be reached other than (1.4) is nonsense.

Furthermore, on account of the fact that the square of the total intensity is

$$I^2 = Q^2 + U^2 + V^2, \tag{1.5}$$

for complete polarization, the expression given for the total intensity,

$$I = \frac{3\sigma_T}{16\pi A} \left\{ \frac{8}{3}\sqrt{\pi}a_{00} + \frac{4}{3}\sqrt{\frac{\pi}{5}}a_{20} \right\}, \tag{1.6}$$

cannot be correct since it does not contain the contribution from (1.4). On the strength of tight coupling, or thermal equilibrium between photons and electrons, it is claimed that photons must have a distribution that 'mirrors' the distribution of electrons. This is hardly possible since photons are bose particles whereas electrons are fermions. Notwithstanding this, an 'immediate consequence' is that the angular dependence of the radiation field can possess both a monopole ($\ell = 0$, with a_{00}), corresponding to temperature fluctuations and dipole component ($\ell = 1$ with a_{10} and a_{11}), corresponding to a Doppler shift caused by a peculiar velocity. The quadrupole is supposedly subsequently produced at decoupling where photons are now free to go their own way.

This does make for good story telling, and it wouldn't be all that bad if the inflationists kept to the same story. But the components of the quadrupole moment, $\ell = 2$, cannot be further decomposed into scalar, $m = 0$, vector, $m = \pm 1$, and tensor, $m = \pm 2$, perturbations [HW97]. And furthermore, temperature differences cannot create polarization so the temperature fluctuations–polarization connection has been severed, or better, never existed.

5. Patchwork on the Standard Model

When Penzias and Wilson first discovered background radiation at one frequency that was spread out uniformly across the sky, no one knew that it had a pure blackbody spectrum. Antichronologically, that prediction was made a decade and a half earlier by Gamow and associates, and the Russians nonchalantly referred to it as 'relic radiation', which was all but obvious that it was there. But the detection of such radiation not only added support to the big bang theory, it also sounded the death kneel of competing theories like Hoyle and colleagues' [FHN00] steady-state theory for such radiation coming from numerous discrete sources would be too contrived to be real.

Subsequently, cracks in the big bang scenario began to appear. One of the gravest was there wasn't enough time for structures to form from the time that the bang supposedly occurred. The standard-bearer of the standard model had to be replaced: Hubble expansion, expressing velocity in terms of distance, gave way to accelerated expansion, the cosmological redshift was replaced by inflation, and the nucleosynthesis of the light elements in the early universe was not sufficient to give the universe its critical density so as to render it flat, and had to give way to dark matter and/or dark energy.

The same pure blackbody spectrum that was found by COBE in 1992, that initially thrilled the big bangers, now became a thorn in what looked like an Achilles' heel: Such homogeneity would not provide the seeds for structure formation. So the blackbody spectrum couldn't be all that pure if inhomogeneities were to form that would be the embryos of the galaxies that we see in the sky.

What has emerged since then is a frantic effort to cover up the deep lacunae in the standard model by transgressing all the known laws of physics. The inflationary model which appeared in the late seventies and early eighties had its origin in de Sitter's teens solution of Einstein's adiabatic equations. Instead of a rigged cosmological constant blowing up the universe, a negative pressure took over its role. And since Einstein's equations are adiabatic there can be no enormous growth in entropy once inflation sets in. How inflation stops, and how the universe recovers from the lethargic and run-down state that inflation left it in, are no clearer today than when inflation was originally proposed.

Where does all this leave us? Not any worse off than the state after inflation has ceased. It is not the experimental method that is at risk, but, rather, the attitude of 'knowing' the results before performing the experiment that is reprehensible. Who writes an experimental paper with the theoretical explanation preceeding the experimental results [col13]?

As I have tried to show in this book, you can't build edifices on quagmires. In the end, it all leads back to Einstein's general relativity, which is much less general than the special case, and certainly much less relative. In short, it is not applicable to the task it was developed for; the universe cannot be modelled as a perfect fluid. The same is even truer of string theory, which cannot boast of a single physical prediction. Multiverse is a sad excuse for not seeking the actual laws that govern physical phenomena. Admittedly, we have not made any theoretical progress in the hundred years that general relativity, and the nearly fifty years that string theory, have been around. It's time for a new start and to wipe the slate clean.

Bibliography

[Ach14]   J. Achenbach. Cosmic smash-up: BICEP2's big bang discovery getting dusted by new satellite data. *Washington Post*, September 22nd 2014.

[Adl]   R. Adler. Interaction between light and sound. *IEEE Spectrum*, 4(5): 42–54, 1967.

[BS62]   D. M. Brink and G. R. Satchler. *Angular Momentum*. Oxford University Press, Oxford, 1962.

[Cha50]   S. Chandrasekhar. *Radiative Transfer*. Oxford University Press, Oxford, 1950.

[col13]   Planck collaboration. Planck 2013 results: XXII. Constraints on inflation. Technical Report arXiv:1303.5082v2, December 5th 2013.

[Com14]   T. Commissariat. BICEP2 gravitational wave bites the dust thanks to new Planck data. *Physics World*, September 22nd 2014.

[Cow14]   R. Cowen. Gravitational wave discovery faces scrutiny. *Nature*, 16 May 2014.

[Dod03]   S. Dodelson. *Modern Cosmology*. Academic Press, San Diego, 2003.

[FHN00]   G. Burbidge F. Hoyle and J. V. Narlikar. *A Different Approach to Cosmology: From a static universe through the big bang towards reality*. Cambridge University Press, Cambridge, 2000.

[Fla96]   T. Van Flandern. Possible new properties of gravity. *Astrophysics & Space Sci.*, 244:249–261, 1996.

[GS13]   F. Graham-Smith. *Unseen Cosmos: The universe in radio*. Oxford University Press, Oxford, 2013.

[HD02]   W. Hu and S. Dodelson. Cosmic Microwave Anisotropies. *Ann. Rev. Astron. and Astrophysics*, pages 1–51, 2002.

[HW97]   W. Hu and M. White. A CMB polarization primer. Technical Report arXiv:astro-ph/9706147v1, June 1997.

[Kos99]   Introduction to microwave background polarization. *New Aston. Rev.*, 43:157–168, 1999.

[Lav11]   B. H. Lavenda. *A New Perspective on Relativity: An odyssey in non-Euclidean geometries*. World Scientific, Singapore, 2011.

[Rod14]   M. C. Rodman. Harvard team's big bang findings called into question. *The Harvard Crimson*, September 24th 2014.

[RT67]   L. S. Rodberg and R. M. Thaler. *Introduction to the Quantum Theory of Scattering*. Academic Press, New York, 1967.

[SB53]   E. Schrödinger and L. Bass. Must the photon mass be zero? *Proc. Roy. Soc. A*, 232:1–6, 1953.

[Shu62]   W. A. Shurcliff. *Polarized Light: Production and use*. Harvard University Press, Cambridge, MA, 1962.

[SZ98]   U. Seljak and M. Zaldarriaga. Polarization of the microwave background: Statistical and physical properties. Technical Report arXiv:astro-ph/9805010v1, May 1998.

[vdH57]   H. C. van de Hulst. *Light Scattering by small articles*. John Wiley & Sons, New York, 1957.

Index